11th International System on Modern Optics and Its Applications (ISMOA 2017)

Journal of Physics: Conference Series
Volume 1057

Bogor, Indonesia
7 – 10 August 2017

ISBN: 978-1-5108-6826-7
ISSN: 1742-6588

Printed from e-media with permission by:

Curran Associates, Inc.
57 Morehouse Lane
Red Hook, NY 12571

Some format issues inherent in the e-media version may also appear in this print version.

Printed by Curran Associates, Inc. (2018)

For permission requests, please contact the Institute of Physics
at the address below.

Institute of Physics
Dirac House, Temple Back
Bristol BS1 6BE UK

Phone: 44 1 17 929 7481
Fax: 44 1 17 920 0979

techtracking@iop.org

Additional copies of this publication are available from:

Curran Associates, Inc.
57 Morehouse Lane
Red Hook, NY 12571 USA
Phone: 845-758-0400
Fax: 845-758-2633
Email: curran@proceedings.com
Web: www.proceedings.com

Table of contents

Volume 1057

11th International Symposium on Modern Optics and Its Applications (ISMOA)

7–10 August 2017, Bogor, Indonesia

Accepted papers received: 4 July 2018
Published online: 27 July 2018

Preface

Papers

PREFACE

The (11th) International Symposium on Modern Optics and Its Applications (ISMOA) is a bi-annual meeting held by the Indonesian Optical Society (InOS)/Himpunan Optika Indonesia (HOI). For 11 times, this symposium has been marked as the largest conference of the kind in Indonesia which not only gather modern optical scientists from all around Indonesia, but also from other regions around the world. ISMOA was initiated by two of the founders of The Indonesian Optical Society (InOS), Prof. Tjia and Dr. Alexander Iskandar in 2001 under the support of KNAW, The Royal Netherland Academy of Arts and Sciences and several colleagues from University of Twente. As time goes by, after the founding of InOS in 2011, ISMOA was co-organized by InOS in collaboration with ITB. For the first time, ISMOA is co-organized by InOS with another group not from ITB, of course still with strong support by InOS members from ITB.

This time, the event was held at the IPB Baranang Siang Campus (Tutorial Session) and IPB International Convention Center (Symposium) in Bogor City West Java Indonesia from the 7th-10th August 2017. The value of this event is to enable scientific communication and interactions among regional scientists and distinguished scientists working in the rapidly growing field of modern optics and novel functional optical materials as well as the associated applications to information and communication technology (ICT), bio-medical sensing, imaging and biophotonics as well as applications to the development of sustainable green energy generation. This event is also designed to offer the opportunity of making direct contacts for the young Indonesian and ASEAN scientists and students with leading researchers abroad, thereby inspiring their research interest and fostering the existing research collaborations as well as extending international research networking for the future.

The invited speakers were:

1. Harith AHMAD (University of Malaya, Malaysia)
2. Carsten ROCKSTHUL (Karlsruhe Institute of Technology, Germany)
3. B.M. Azizur RAHMAN (City University of London, United Kingdom)
4. Kurt HINGERL (Johannes Kepler University, Linz, Austria)
5. Yong-Hee LEE (Korea Institute for Advanced Study, Korea)
6. Clarck G. LI (University of Sydney, Australia)
7. David SAMPSON (University of Western Australia, Australia)
8. Kaoru TAMADA (University of Kyushu, Japan)

The symposium event resulted in 12 papers publised in the SCOPUS/Thompson-Reuters Indexed proceeding of the Institute of Physics, Journal of Physics Conference Series The acceptance rate of the papers were 100% since we conducted earlier peer review on the extended abstract submitted by the participants before attending the conference. We thank IOP for their professionalism and support of publishing the works in this memorable event.

Bogor, 29 March 2018
ISMOA 2017 Chairman

Prof. Dr. Husin Alatas

ISMOA Proceeding Chair

Dr. rer. nat. Hendradi Hardhienata

IOP Conf. Series: Journal of Physics: Conf. Series **1057** (2018) 011002 doi:10.1088/1742-6596/1057/1/011002

Reviewer List for ISMOA 2017

No	Reviewer Name	Email	Institution
1	Prof. Dr. Tjia May On	fismots@fi.itb.ac.id	Department of Physics, Institut Teknologi Bandung, INDONESIA
2	Dr. Alexander A. Iskandar	iskandar@fi.itb.ac.id	Department of Biology, Institut Teknologi Bandung, MALAYSIA
3	Prof. Dr. Husin Alatas	alatas@ipb.ac.id	Department of Physics, Bogor Agricultural University, INDONESIA
4	Dr. rer. nat. Hendradi Hardhienata	hendradi@ipb.ac.id	Department of Physics, Bogor Agricultural University, INDONESIA

Peer review statement

All papers published in this volume of *Journal of Physics: Conference Series* have been peer reviewed through processes administered by the proceedings Editors. Reviews were conducted by expert referees to the professional and scientific standards expected of a proceedings journal published by IOP Publishing.

IOP Conf. Series: Journal of Physics: Conf. Series **1057** (2018) 012001 doi:10.1088/1742-6596/1057/1/012001

Isotropic and Anisotropic Parts for the Susceptibility Tensor Calculated Using Simplified Bond-Hyperpolarizability Model

A Alejo-Molina[1] and H Hardhienata[2]

[1] CONACYT Research Fellow–Center for Research in Engineering and Applied Science (CIICAp), Institute for Research in Pure and Applied Science (IICBA), UAEM Cuernavaca, Mor. 62209, Mexico
[2] Department of Physics, Bogor Agricultural University, Jl. Meranti, Gedung Wing S, Kampus IPB Darmaga, Bogor 16680, Jawa Barat, Indonesia

E-mail: adalberto.alejo@uaem.mx

Abstract. In this paper we discuss the importance of separating the susceptibility tensor in the isotropic and anisotropic parts, we show how to do that and what is the contribution from each part to the nonlinear polarizability. Specific examples are given for second harmonic generation in the surface of Si(001) and for third harmonic generation in bulk silicon along (001) direction. Our calculations are performed using simplified bond-hyperpolarizability model.

1. Introduction

Nonlinear optics is a favorite method in the investigation of surface properties of semiconductor materials due to the possibility to probe materials at room temperature, is relatively nondestructive, can be performed in non vacuum condition, but mainly because it is surface sensitive. One of the interesting issues in probing surfaces of semiconductors using nonlinear optics (e.g. higher harmonics) is to identify and separate the nonlinear sources that are generated from the surface and the bulk of the material. In doing so analysis of the nonlinear susceptibility tensor becomes an important part because it is directly related to the symmetry of the material that produces the nonlinear sources.

In a classic paper by Sipe and coworkers [1], they split the tensor describing the nonlocal response for the nonlinear polarization in the isotropic and the anisotropic parts. In general a tensor can be separated in the symmetric and the antisymmetric part or in the isotropic and the anisotropic part [2]. The latter procedure is our interest here, because for second harmonic generation (SHG) on crystals for the surface or the bulk, a common technique is to rotate the crystal around the normal (usually labeled *z*-axis). This technique is known as Rotational Anisotropy of the SHG (RA-SHG) and the experimental results can be well described using Simplified Bond-Hyperpolarizability Model [3]. Then the intensity of the SH (second harmonic) signal is a function of the azimuthal angle and change sinusoidally with an integer number of periods for a complete revolution of the crystal.

Describing harmonic generation with SBHM, generates a susceptibility tensor. This is, the polarization is given by [4]

$$\vec{P} = \frac{1}{V}\sum_j \alpha_{1j}\left[\hat{b}_j \otimes \hat{b}_j\right]\cdot \vec{E} + \frac{1}{V}\sum_j \alpha_{2j}\left[\hat{b}_j \otimes \hat{b}_j \otimes \hat{b}_j\right]\cdot\cdot \vec{E}\otimes\vec{E}$$
$$+\frac{1}{V}\sum_j \alpha_{3j}\left[\hat{b}_j \otimes \hat{b}_j \otimes \hat{b}_j \otimes \hat{b}_j\right]\cdots \vec{E}\otimes\vec{E}\otimes\vec{E} + \ldots \quad (1)$$
$$= \chi^{(1)}\vec{E} + \vec{\vec{\chi}}^{(2)}\vec{E}\vec{E} + \vec{\vec{\chi}}^{(3)}\vec{E}\vec{E}\vec{E} + \ldots$$

where V is the volume, α_{1j} are the linear polarizabilities, α_{2j} and α_{3j} are the first and second order hyperpolarizabilities and $\hat{b}_j$ are the unit vectors in the direction of the atomic bonds; whereas $\vec{E}(\omega)$ is the electric field. Incident polarization can be defined for the electric field (typically *s*- or *p*-polarization) and even a DC field can be modeled for example for describing electric field induced second harmonic (EFISH) generation [4]. However, we are going to center our discussion about the tensorial nature of the susceptibility.

For a second-order nonlinear phenomena, there is a third-rank tensor which corresponds with the susceptibility represented by $\vec{\vec{\chi}}^{(2)}$. This tensor, in general has 27 elements and can be represented by [6]

$$\vec{\vec{\chi}}^{(2)} = \begin{pmatrix} \begin{pmatrix} \chi_{111} & \chi_{121} & \chi_{131} \\ \chi_{112} & \chi_{122} & \chi_{132} \\ \chi_{113} & \chi_{123} & \chi_{133} \end{pmatrix} \\ \begin{pmatrix} \chi_{211} & \chi_{221} & \chi_{231} \\ \chi_{212} & \chi_{222} & \chi_{232} \\ \chi_{213} & \chi_{223} & \chi_{233} \end{pmatrix} \\ \begin{pmatrix} \chi_{311} & \chi_{321} & \chi_{331} \\ \chi_{312} & \chi_{322} & \chi_{332} \\ \chi_{313} & \chi_{323} & \chi_{333} \end{pmatrix} \end{pmatrix} \quad (2)$$

where in the component representation of the tensor $\chi^{(2)}_{ijk}$, the first index "i" corresponds to the rows in the main external matrix, whereas the column has no index associated. Thus, all the elements in the first row of the inner 3×3 matrix have χ_{1jk} indices, whereas for the second row it will be χ_{2jk} and so on. The second and third indices "j" and "k" will correspond to the usual way of labeling a 3×3 matrix, they are the rows and columns, respectively, in the inner 3×3 matrix.

On the other hand, a third-order nonlinear phenomena will be described by a fourth-rank susceptibility tensor. A general fourth-rank tensor, can be represented as a 3×3 matrix which also has a 3×3 matrices as elements, this is [6]

$$\overleftrightarrow{\chi}^{(3)} = \begin{pmatrix} \begin{pmatrix} \chi_{1111} & \chi_{1112} & \chi_{1113} \\ \chi_{1121} & \chi_{1122} & \chi_{1123} \\ \chi_{1131} & \chi_{1132} & \chi_{1133} \end{pmatrix} & \begin{pmatrix} \chi_{1211} & \chi_{1212} & \chi_{1213} \\ \chi_{1221} & \chi_{1222} & \chi_{1223} \\ \chi_{1231} & \chi_{1232} & \chi_{1233} \end{pmatrix} & \begin{pmatrix} \chi_{1311} & \chi_{1312} & \chi_{1313} \\ \chi_{1321} & \chi_{1322} & \chi_{1323} \\ \chi_{1331} & \chi_{1332} & \chi_{1333} \end{pmatrix} \\ \begin{pmatrix} \chi_{2111} & \chi_{2112} & \chi_{2113} \\ \chi_{2121} & \chi_{2122} & \chi_{2123} \\ \chi_{2131} & \chi_{2132} & \chi_{2133} \end{pmatrix} & \begin{pmatrix} \chi_{2211} & \chi_{2212} & \chi_{2213} \\ \chi_{2221} & \chi_{2222} & \chi_{2223} \\ \chi_{2231} & \chi_{2232} & \chi_{2233} \end{pmatrix} & \begin{pmatrix} \chi_{2311} & \chi_{2312} & \chi_{2313} \\ \chi_{2321} & \chi_{2322} & \chi_{2323} \\ \chi_{2331} & \chi_{2332} & \chi_{2333} \end{pmatrix} \\ \begin{pmatrix} \chi_{3111} & \chi_{3112} & \chi_{3113} \\ \chi_{3121} & \chi_{3122} & \chi_{3123} \\ \chi_{3131} & \chi_{3132} & \chi_{3133} \end{pmatrix} & \begin{pmatrix} \chi_{3211} & \chi_{3212} & \chi_{3213} \\ \chi_{3221} & \chi_{3222} & \chi_{3223} \\ \chi_{3231} & \chi_{3232} & \chi_{3233} \end{pmatrix} & \begin{pmatrix} \chi_{3311} & \chi_{3312} & \chi_{3313} \\ \chi_{3321} & \chi_{3322} & \chi_{3323} \\ \chi_{3331} & \chi_{3332} & \chi_{3333} \end{pmatrix} \end{pmatrix}. \quad (3)$$

There are 81 elements, which are labeled as follows. The first index "*i*" in $\chi^{(3)}_{ijkl}$ corresponds to the rows and the second index "*j*" to the columns in the main matrix (the external one). In the same way, the indices "*k*" and "*l*" will correspond to the usual way of labeling a 3×3 matrix, namely, the rows and columns in the inner 3×3 matrix, respectively.

Fortunately, as is well-known, many of the tensorial elements in equations (2) and (3) are zero for crystals with high symmetry. This is because the susceptibility tensor of the crystal should be the same tensor after a transformation due to some intrinsic symmetry in the crystal and this is the way in which non-zero elements in the tensor are calculated [6]; in Physics this is known as Neumann's principle [7]. Mathematically, a transformation by a rotation or by another symmetry operation applied to the susceptibility *n*th-rank tensor is calculated by

$$\chi'_{x_1x_2x_3\ldots x_n} = R_{x_1y_1}R_{x_2y_2}R_{x_3y_3}R_{x_4y_4}\ldots R_{x_ny_n}\chi_{y_1y_2y_3\ldots y_n}, \quad (4)$$

where, as it is mentioned before, the R_{xy} is a matrix defining a symmetry operation or could be a general rotation for an arbitrary angle θ, around *z*-axis. Note that basically there is a matrix contracting each index in the tensor. For a general rotation $\chi'_{x_1x_2x_3\ldots x_n} = \chi'_{x_1x_2x_3\ldots x_n}(\theta)$, and then it can be separated in isotropic and anisotropic part. Isotropic part is invariant under rotations and the anisotropic part should have all the information about the azimuthal dependence. Thus something is isotropic when it does not change in function of the direction.

A third-rank tensor is isotropic under rotations when it has the form

$$\overleftrightarrow{\varepsilon} = \begin{pmatrix} \begin{pmatrix} 0 & 0 & 0 \\ 0 & 0 & -1 \\ 0 & 1 & 0 \end{pmatrix} \\ \begin{pmatrix} 0 & 0 & 1 \\ 0 & 0 & 0 \\ -1 & 0 & 0 \end{pmatrix} \\ \begin{pmatrix} 0 & -1 & 0 \\ 1 & 0 & 0 \\ 0 & 0 & 0 \end{pmatrix} \end{pmatrix} \quad (5)$$

which is nothing else than the Levi-Civita tensor ε_{ijk} [9]. This can be verified by the procedure followed in Reference [7] applied to the tensor given in equation (2), where for anticlockwise rotations around all

IOP Conf. Series: Journal of Physics: Conf. Series **1057** (2018) 012001 doi:10.1088/1742-6596/1057/1/012001

the Cartesian axes x, y and z the result is $\chi_{123}\varepsilon_{ijk}$ and in the case of clockwise rotation there is only a minus one multiplying the last result.

In the same way, it is possible to find a fourth-rank tensor which is invariant under rotations in any Cartesian axis. The fourth-rank isotropic tensor has the general form [2, 9]:

$$\vec{I}=\begin{pmatrix} \begin{pmatrix} \alpha+\beta+\gamma & 0 & 0\\ 0 & \alpha & 0\\ 0 & 0 & \alpha \end{pmatrix} & \begin{pmatrix} 0 & \beta & 0\\ \gamma & 0 & 0\\ 0 & 0 & 0 \end{pmatrix} & \begin{pmatrix} 0 & 0 & \beta\\ 0 & 0 & 0\\ \gamma & 0 & 0 \end{pmatrix}\\ \begin{pmatrix} 0 & \gamma & 0\\ \beta & 0 & 0\\ 0 & 0 & 0 \end{pmatrix} & \begin{pmatrix} \alpha & 0 & 0\\ 0 & \alpha+\beta+\gamma & 0\\ 0 & 0 & \alpha \end{pmatrix} & \begin{pmatrix} 0 & 0 & 0\\ 0 & 0 & \beta\\ 0 & \gamma & 0 \end{pmatrix}\\ \begin{pmatrix} 0 & 0 & \gamma\\ 0 & 0 & 0\\ \beta & 0 & 0 \end{pmatrix} & \begin{pmatrix} 0 & 0 & 0\\ 0 & 0 & \gamma\\ 0 & \beta & 0 \end{pmatrix} & \begin{pmatrix} \alpha & 0 & 0\\ 0 & \alpha & 0\\ 0 & 0 & \alpha+\beta+\gamma \end{pmatrix} \end{pmatrix}. \tag{6}$$

where α, β and γ are arbitrary constants (scalars). This can be checked using equation (4) for a rotation matrix defined for any Cartesian axis, after transforming the tensor the result is going to be the same tensor given in equation (6). As can be seen, isotropic tensors are more complicated when the rank is higher.

In the next sections, specific examples of the isotropic and the anisotropic parts of the susceptibility tensor will be given for third- and fourth-rank tensors generated through SBHM. Finally, the conclusions and remarks of this work are given.

2. SBHM second-order nonlinear phenomena

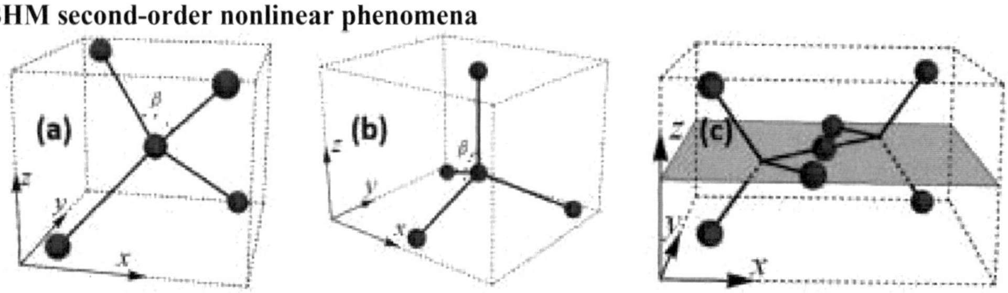

Figure 1. Tetrahedral elements representing the facets (a) Si(001) (b) Si(111) and (c) Si(110)

First, we are going to discuss SHG in the surface on silicon with the facet (001). The bond vectors in this case are

$$\vec{b}_1=-\frac{1}{\sqrt{2}}\begin{pmatrix} \sin\frac{\beta}{2}\\ \sin\frac{\beta}{2}\\ \sqrt{2}\cos\frac{\beta}{2} \end{pmatrix},\ \vec{b}_2=\frac{1}{\sqrt{2}}\begin{pmatrix} \sin\frac{\beta}{2}\\ \sin\frac{\beta}{2}\\ -\sqrt{2}\cos\frac{\beta}{2} \end{pmatrix},\ \vec{b}_3=\frac{1}{\sqrt{2}}\begin{pmatrix} -\sin\frac{\beta}{2}\\ \sin\frac{\beta}{2}\\ \sqrt{2}\cos\frac{\beta}{2} \end{pmatrix},\ \vec{b}_4=\frac{1}{\sqrt{2}}\begin{pmatrix} \sin\frac{\beta}{2}\\ -\sin\frac{\beta}{2}\\ \sqrt{2}\cos\frac{\beta}{2} \end{pmatrix}, \tag{7}$$

which are plotted schematically in figure 1 (a). Therefore, the susceptibility tensor is calculated using the second term in equation (1)

$$\vec{\chi}^{(2)} = \frac{1}{V}\sum_{j=1}^{4}\alpha_{2j}\left\{\left[\mathbf{R}^{(z)}(\phi)\cdot\hat{b}_j\right]\otimes\left[\mathbf{R}^{(z)}(\phi)\cdot\hat{b}_j\right]\otimes\left[\mathbf{R}^{(z)}(\phi)\cdot\hat{b}_j\right]\right\} \tag{8}$$

where α_{2j} has two different values in this case, depending if the bond vector point "up" or "down" and they are labeled α_u and α_d. Also, $\mathbf{R}^{(z)}(\phi)$ is the rotation matrix around *z*-axis and it is just for the case that the $\hat{b}_j$s are chosen in a different system of reference with only the *z*-axis conserved, otherwise it is only needed to take $\phi = 0$. Thus, equation (8) yields

$$\vec{\chi}^{(2)}(001) = \begin{pmatrix} \begin{pmatrix} 0 & 0 & 2S\left[\alpha_u\cos^2\phi + \alpha_d\sin^2\phi\right] \\ 0 & 0 & (\alpha_u - \alpha_d)S\sin 2\phi \\ 2S\left[\alpha_u\cos^2\phi + \alpha_d\sin^2\phi\right] & (\alpha_u - \alpha_d)S\sin 2\phi & 0 \end{pmatrix} \\ \begin{pmatrix} 0 & 0 & (\alpha_u - \alpha_d)S\sin 2\phi \\ 0 & 0 & 2S\left[\alpha_u\sin^2\phi + \alpha_d\cos^2\phi\right] \\ (\alpha_u - \alpha_d)S\sin 2\phi & 2S\left[\alpha_u\sin^2\phi + \alpha_d\cos^2\phi\right] & 0 \end{pmatrix} \\ \begin{pmatrix} 2S\left[\alpha_u\cos^2\phi + \alpha_d\sin^2\phi\right] & (\alpha_u - \alpha_d)S\sin 2\phi & 0 \\ (\alpha_u - \alpha_d)S\sin 2\phi & 2S\left[\alpha_u\sin^2\phi + \alpha_d\cos^2\phi\right] & 0 \\ 0 & 0 & 2(\alpha_u + \alpha_d)\cos^2(\beta/2) \end{pmatrix} \end{pmatrix}, \tag{9}$$

where (001) in $\vec{\chi}^{(2)}$ is just to remember in which direction is this tensor defined, $S = \sin\beta\sin(\beta/2)/2$ and in particular for silicon $\beta = 2\arccos\left(1/\sqrt{3}\right) \approx 109.47°$. Now by direct comparison between this tensor equation (9) and the isotropic one given in equation (5) it is possible to separate the isotropic and the anisotropic parts:

$$\vec{\chi}^{(2)}(001) = \vec{\chi}^{(2)}_{ISO} + \vec{\chi}^{(2)}_{ANI} \tag{10}$$

whereas the isotropic part is explicitly

$$\vec{\chi}^{(2)}_{ISO} = (\alpha_u - \alpha_d)S\sin 2\phi \begin{pmatrix} \begin{pmatrix} 0 & 0 & 0 \\ 0 & 0 & -1 \\ 0 & 1 & 0 \end{pmatrix} \\ \begin{pmatrix} 0 & 0 & 1 \\ 0 & 0 & 0 \\ -1 & 0 & 0 \end{pmatrix} \\ \begin{pmatrix} 0 & -1 & 0 \\ 1 & 0 & 0 \\ 0 & 0 & 0 \end{pmatrix} \end{pmatrix}, \tag{11}$$

and the anisotropic part is

IOP Conf. Series: Journal of Physics: Conf. Series **1057** (2018) 012001 doi:10.1088/1742-6596/1057/1/012001

$$\vec{\chi}^{(2)}_{ANI} = \begin{pmatrix} \begin{pmatrix} 0 & 0 & 2S\left[\alpha_u \cos^2\phi + \alpha_d \sin^2\phi\right] \\ 0 & 0 & 2(\alpha_u - \alpha_d) S \sin 2\phi \\ 2S\left[\alpha_u \cos^2\phi + \alpha_d \sin^2\phi\right] & 0 & 0 \end{pmatrix} \\ \begin{pmatrix} 0 & 0 & 0 \\ 0 & 0 & 2S\left[\alpha_u \sin^2\phi + \alpha_d \cos^2\phi\right] \\ 2(\alpha_u - \alpha_d) S \sin 2\phi & 2S\left[\alpha_u \sin^2\phi + \alpha_d \cos^2\phi\right] & 0 \end{pmatrix} \\ \begin{pmatrix} 2S\left[\alpha_u \cos^2\phi + \alpha_d \sin^2\phi\right] & 2(\alpha_u - \alpha_d) S \sin 2\phi & 0 \\ 0 & 2S\left[\alpha_u \sin^2\phi + \alpha_d \cos^2\phi\right] & 0 \\ 0 & 0 & 2(\alpha_u + \alpha_d)\cos^2(\beta/2) \end{pmatrix} \end{pmatrix}. \quad (12)$$

It is clear from equation (11) that for some elections of the system of reference for the bond vectors there is not an isotropic part of the SBHM susceptibility tensor for facet (001). Please, note that the angle ϕ, is for choosing some system of reference for the atomic bonds and if we apply a rotation by and angle θ to the tensor in equation (12) it will be invariant under this general rotation, *i.e.* the rotation will give the same tensor after this operation and will not have any dependence with the angle θ. However, if a general rotation is applied to the tensor given by equation (12), it will change and will have and explicit dependence in the angle θ.

In order to compare both results, we are going to find the second harmonic resulting polarization for the particular case of s-polarization as the fundamental excitation. Thus the electric field only will have components in the y direction $\vec{E}_s = \left(0, E_y(\omega), 0\right)$ and after contracting with the full tensor in equation (9):

$$\vec{P}_s(2\omega) = 2SE_y^2\left(0, 0, \alpha_d \cos^2\phi + \alpha_u \sin^2\phi\right). \quad (13)$$

Now, for the case in which the tensor was split in the isotropic and anisotropic parts:

$$\vec{P}_s(2\omega) = \left[\vec{\chi}^{(2)}_{ISO} + \vec{\chi}^{(2)}_{ANI}\right] \cdot\cdot \vec{E} \otimes \vec{E} = 0 + 2SE_y^2\left(0, 0, \alpha_d \cos^2\phi + \alpha_u \sin^2\phi\right), \quad (14)$$

as should be, isotropic part is just a constant contribution to the total intensity of the SHG and in this particular case it is zero; whereas the anisotropic part has all the information about the azimuthal dependence of the intensity. The same is true for the p-polarization, but we do not show it explicitly here.

In the case of the facet Si(111), the isotropic and anisotropic parts of the susceptibility tensor can be obtained from the correspondent ones for the Si(001) facet. It is only needed to rotate the anisotropic part of the tensor in equation (12) and then add the result with the tensor in equation (11), because the isotropic part is by definition invariant under rotations. As the orientation is different we need to rotate the crystal to a system of reference with the z-axis perpendicular to the plane (111). For doing this, two rotations must be performed [10], the first one is around the z-axis by an angle of $\pi/4$ (clockwise); then in the new system of reference the second rotation will be around x-axis, by $\beta/2$ (counterclockwise), where β is the angle between the bonds and it is equal to 109.47° for silicon. The final configuration is shown in figure 1 (b). This procedure of rotating the previous result, can be applied too to facet Si(011), shown in figure 1 (c), but the angles and axis are different that in the previous case [10]. Other procedure is to calculate the susceptibility tensor with SBHM [5, 10, 11] starting with the bonds vectors in the adequate direction and then to separate by inspection the isotropic and anisotropic parts.

3. SBHM third-order nonlinear phenomena

For third-order nonlinear phenomena we are going to do the same separation of the isotropic and anisotropic parts of the susceptibility fourth-rank tensor but only for a particular nonlinear effect. In this case, we are going to discuss third harmonic generation for Si(001) only. The SBHM equation describing it, is the last summand in equation (1). Then, the susceptibility tensor is

$$\vec{\vec{\chi}}^{(3)}_{SBHM} = \alpha_3 \begin{pmatrix} \begin{pmatrix} [3+\cos(4\phi)]\sin^4\left(\frac{\beta}{2}\right) & \sin(4\phi)\sin^4\left(\frac{\beta}{2}\right) & 0 \\ \sin(4\phi)\sin^4\left(\frac{\beta}{2}\right) & 2\sin^2(2\phi)\sin^4\left(\frac{\beta}{2}\right) & 0 \\ 0 & 0 & \sin^2\beta \end{pmatrix} & \begin{pmatrix} \sin(4\phi)\sin^4\left(\frac{\beta}{2}\right) & 2\sin^2(2\phi)\sin^4\left(\frac{\beta}{2}\right) & 0 \\ 2\sin^2(2\phi)\sin^4\left(\frac{\beta}{2}\right) & -\sin(4\phi)\sin^4\left(\frac{\beta}{2}\right) & 0 \\ 0 & 0 & 0 \end{pmatrix} & \begin{pmatrix} 0 & 0 & \sin^2\beta \\ 0 & 0 & 0 \\ \sin^2\beta & 0 & 0 \end{pmatrix} \\ \begin{pmatrix} \sin(4\phi)\sin^4\left(\frac{\beta}{2}\right) & 2\sin^2(2\phi)\sin^4\left(\frac{\beta}{2}\right) & 0 \\ 2\sin^2(2\phi)\sin^4\left(\frac{\beta}{2}\right) & -\sin(4\phi)\sin^4\left(\frac{\beta}{2}\right) & 0 \\ 0 & 0 & 0 \end{pmatrix} & \begin{pmatrix} 2\sin^2(2\phi)\sin^4\left(\frac{\beta}{2}\right) & -\sin(4\phi)\sin^4\left(\frac{\beta}{2}\right) & 0 \\ -\sin(4\phi)\sin^4\left(\frac{\beta}{2}\right) & [3+\cos(4\phi)]\sin^4\left(\frac{\beta}{2}\right) & 0 \\ 0 & 0 & \sin^2\beta \end{pmatrix} & \begin{pmatrix} 0 & 0 & 0 \\ 0 & 0 & \sin^2\beta \\ 0 & \sin^2\beta & 0 \end{pmatrix} \\ \begin{pmatrix} 0 & 0 & \sin^2\beta \\ 0 & 0 & 0 \\ \sin^2\beta & 0 & 0 \end{pmatrix} & \begin{pmatrix} 0 & 0 & 0 \\ 0 & 0 & \sin^2\beta \\ 0 & \sin^2\beta & 0 \end{pmatrix} & \begin{pmatrix} \sin^2\beta & 0 & 0 \\ 0 & \sin^2\beta & 0 \\ 0 & 0 & 8\cos^4\left(\frac{\beta}{2}\right) \end{pmatrix} \end{pmatrix}. \tag{15}$$

There is only one hyperpolarizability because this is a bulk effect [10]. In this case, it is also possible to separate in general the isotropic and anisoptropic parts of the tensor, but an easy example is going to be shown. For instance, if the azimuthal angle $\phi = \pi/4$, the general tensor given by equation (15) yields

$$\vec{\vec{\chi}}^{(3)}_{SBHM}(\phi=\pi/4) = \alpha_3 \begin{pmatrix} \begin{pmatrix} 2\sin^4\left(\frac{\beta}{2}\right) & 0 & 0 \\ 0 & 2\sin^4\left(\frac{\beta}{2}\right) & 0 \\ 0 & 0 & \sin^2\beta \end{pmatrix} & \begin{pmatrix} 0 & 2\sin^4\left(\frac{\beta}{2}\right) & 0 \\ 2\sin^4\left(\frac{\beta}{2}\right) & 0 & 0 \\ 0 & 0 & 0 \end{pmatrix} & \begin{pmatrix} 0 & 0 & \sin^2\beta \\ 0 & 0 & 0 \\ \sin^2\beta & 0 & 0 \end{pmatrix} \\ \begin{pmatrix} 0 & 2\sin^4\left(\frac{\beta}{2}\right) & 0 \\ 2\sin^4\left(\frac{\beta}{2}\right) & 0 & 0 \\ 0 & 0 & 0 \end{pmatrix} & \begin{pmatrix} 2\sin^4\left(\frac{\beta}{2}\right) & 0 & 0 \\ 0 & 2\sin^4\left(\frac{\beta}{2}\right) & 0 \\ 0 & 0 & \sin^2\beta \end{pmatrix} & \begin{pmatrix} 0 & 0 & 0 \\ 0 & 0 & \sin^2\beta \\ 0 & \sin^2\beta & 0 \end{pmatrix} \\ \begin{pmatrix} 0 & 0 & \sin^2\beta \\ 0 & 0 & 0 \\ \sin^2\beta & 0 & 0 \end{pmatrix} & \begin{pmatrix} 0 & 0 & 0 \\ 0 & 0 & \sin^2\beta \\ 0 & \sin^2\beta & 0 \end{pmatrix} & \begin{pmatrix} \sin^2\beta & 0 & 0 \\ 0 & \sin^2\beta & 0 \\ 0 & 0 & 8\cos^4\left(\frac{\beta}{2}\right) \end{pmatrix} \end{pmatrix} \tag{16}$$

which we can compare with the general form of the fourth-rank isotropic tensor in equation (6) and after some algebraic and trigonometric manipulation it is easy to separate the isotropic part as

$$\vec{\chi}_{ISO}{}^{(3)}=\alpha_3\begin{pmatrix}\begin{pmatrix}3\sin^4\left(\frac{\beta}{2}\right)&0&0\\0&\sin^4\left(\frac{\beta}{2}\right)&0\\0&0&\sin^2\beta\end{pmatrix}&\begin{pmatrix}0&\sin^4\left(\frac{\beta}{2}\right)&0\\\sin^4\left(\frac{\beta}{2}\right)&0&0\\0&0&0\end{pmatrix}&\begin{pmatrix}0&0&\sin^2\beta\\0&0&0\\\sin^2\beta&0&0\end{pmatrix}\\\begin{pmatrix}0&\sin^4\left(\frac{\beta}{2}\right)&0\\\sin^4\left(\frac{\beta}{2}\right)&0&0\\0&0&0\end{pmatrix}&\begin{pmatrix}\sin^4\left(\frac{\beta}{2}\right)&0&0\\0&3\sin^4\left(\frac{\beta}{2}\right)&0\\0&0&\sin^2\beta\end{pmatrix}&\begin{pmatrix}0&0&0\\0&0&\sin^2\beta\\0&\sin^2\beta&0\end{pmatrix}\\\begin{pmatrix}0&0&\sin^2\beta\\0&0&0\\\sin^2\beta&0&0\end{pmatrix}&\begin{pmatrix}0&0&0\\0&0&\sin^2\beta\\0&\sin^2\beta&0\end{pmatrix}&\begin{pmatrix}\sin^2\beta&0&0\\0&\sin^2\beta&0\\0&0&3\cos^4\left(\frac{\beta}{2}\right)\end{pmatrix}\end{pmatrix} \tag{17}$$

whereas the anisotropic parts is then

$$\vec{\chi}_{ANI}{}^{(3)}=\alpha_3\begin{pmatrix}\begin{pmatrix}1-\cos\beta-\frac{7}{2}\sin^2\beta&0&0\\0&1-\cos\beta-\frac{3}{2}\sin^2\beta&0\\0&0&0\end{pmatrix}&\begin{pmatrix}0&1-\cos\beta-\frac{3}{2}\sin^2\beta&0\\1-\cos\beta-\frac{3}{2}\sin^2\beta&0&0\\0&0&0\end{pmatrix}&\begin{pmatrix}0&0&0\\0&0&0\\0&0&0\end{pmatrix}\\\begin{pmatrix}0&1-\cos\beta-\frac{3}{2}\sin^2\beta&0\\1-\cos\beta-\frac{3}{2}\sin^2\beta&0&0\\0&0&0\end{pmatrix}&\begin{pmatrix}1-\cos\beta-\frac{3}{2}\sin^2\beta&0&0\\0&1-\cos\beta-\frac{7}{2}\sin^2\beta&0\\0&0&0\end{pmatrix}&\begin{pmatrix}0&0&0\\0&0&0\\0&0&0\end{pmatrix}\\\begin{pmatrix}0&0&0\\0&0&0\\0&0&0\end{pmatrix}&\begin{pmatrix}0&0&0\\0&0&0\\0&0&0\end{pmatrix}&\begin{pmatrix}\sin^2\beta&0&0\\0&\sin^2\beta&0\\0&0&4+4\cos\beta-5\sin^2\beta\end{pmatrix}\end{pmatrix}. \tag{18}$$

In contrast with the previous case, the contribution of the isotropic part of the tensor equation (17), after contracting with the electric fields describing s- and p- polarization, is not zero but a constant vector, here we only give explicitly the result for s-polarization

$$\vec{P}_{s,ISO}(3\omega)=\vec{\chi}^{(2)}_{ISO}\cdots\vec{E}\otimes\vec{E}\otimes\vec{E}=3E_y^3\alpha\sin^2\beta(0,1,0,). \tag{19}$$

Here this contribution of the isotropic part is not only an addition to the intensity of the total polarization because the part of the polarization coming from the anisotropic part of the tensor equation (18) also has component in y direction, this is

$$\vec{P}_{s,ANI}(3\omega)=\vec{\chi}^{(2)}_{ANI}\cdots\vec{E}\otimes\vec{E}\otimes\vec{E}=-E_y^3\alpha(5+7\cos\beta)\sin^2\left(\frac{\beta}{2}\right)(0,1,0,) \tag{20}$$

which generates cross terms when the intensity is calculated. Moreover, the same happens with the p-polarization.

4. Conclusions

We have shown that it is always possible to separate the susceptibility tensor, into its isotropic and anisotropic parts. For the case of second harmonic generation tested with *s*- and *p*-polarizations, which is performed by calculating the tensor using SBHM. The separation of isotropic and anisotropic parts, does not offer new insight for the nonlinear polarization, because the contribution of the isotropic part to the nonlinear polarization is zero. This is different for third harmonic generation where the isotropic part has a contribution to the polarization that is not only the addition of a constant to the total intensity but it has a contribution to the angular dependence due to the presence of cross terms in the intensity.

Acknowledgments

A. A. M. thanks to the Mexican National Council for Science and Technology (CONACyT) for financial support during this research (under contract CB-2015-01/254617 “Theoretical Description of Nonlinear Optical Interactions for Surfaces and Bulk, Using Simplified Bond Hyperpolarizability Model and Group Theory”). H.H would like to acknowledge funding from the “Penelitian Unggulan Perguruan Tinggi 2017 No: 011/SP2H/LT/DRPM/IV/2017”.

References

[1] Sipe J E, Moss D J and van Driel H M 1987 Phenomenological theory of optical second- and third harmonic generation from cubic centrosymmetric crystals *Phys. Rev.* B **35** 1129–41
[2] Jeffreys H 1931 *Cartesian Tensors* (London: Cambridge), pp 5, 66–70
[3] Hardhienata H, Alejo-Molina A, Reitböck C, Prylepa A, Stifter D and Hingerl K 2016 Bulk dipolar contribution to second-harmonic generation in zincblende *J. Opt. Soc. Am.* B **33** 195–201
[4] Powell G D, Wang J F and Aspnes D E 2002 Simplified bond-hyperpolarizability model of second harmonic generation *Phys. Rev.* B **65** 205320
[5] Alejo-Molina A, Hardhienata H, Márquez-Aguilar P A and Hingerl K 2017 Facet-dependent electric-field-induced second harmonic generation in silicon and zincblende *J. Opt. Soc. Am.* B **34** 1107–14
[6] Powell R C 2010 *Symmetry, Group Theory, and the Physical Properties of Crystals* (Heidelberg: Springer) Vol. 824
[7] Jatirian-Foltides E S, Escobedo-Alatorre J J, Márquez-Aguilar P A, Hardhienata H, Hingerl K and Alejo-Molina A 2016 About the calculation of the second-order susceptibility $\chi^{(2)}$ tensorial elements for crystals using group theory *Rev. Mex. Fis.* E **62** 5–13
[8] Nye J F 2006 *Physical Properties of Crystals, Their Representations by Tensors and Matrices* (New York: Clarendon)
[9] Arfken G B, Weber H J, Harris F E 2013 *Mathematical Methods for Physicists* (New York: Academic) pp 216
[10] Alejo-Molina A, Hingerl K and Hardhienata H 2015 Model of third harmonic generation and electric field induced optical second harmonic using simplified bond-hyperpolarizability model *J. Opt. Soc. Am.* B **32** 562–70
[11] Alejo-Molina A, Hardhienata H and Hingerl K 2014 Simplified bond-hyperpolarizability model of second harmonic generation, group theory, and Neumann's principle *J. Opt. Soc. Am.* B **31** 526–33

IOP Conf. Series: Journal of Physics: Conf. Series **1057** (2018) 012002 doi:10.1088/1742-6596/1057/1/012002

Detection of Dye Molecules Adsorbed in a Mesoporous Layer by Surface Plasmon Resonance Spectroscopy and its Comparison with Simulation Results

Siti Chalimah, Jalu Setiya Pradana, and Rahmat Hidayat

Physics of Magnetism and Photonics Research Division, Faculty of Mathematics and Natural Sciences, Bandung Institute of Technology, Indonesia

E-mail: sitichalimah31@gmail.com, rahmat@fi.itb.ac.id

Abstract. We report in this paper, the utilization of Surface Plasmon Resonance (SPR) spectroscopy to distinct the adsorption type of dye molecules into an Aluminium-doped Zinc Oxide (AZO) mesoporous layer. This was performed in a Kretschmann configuration by measuring the SPR spectra at various incident angles. The SPR spectral dips were then compared with the simulation results calculated on the basis of the transfer-matrix method. There is an agreement between the simulation results and experimental results, which confirm two different kinds of dye adsorption. The first type is pure dye adsorption into the mesoporous layer, while the second type is dye adsorption accompanied by dye deposition on the mesoporous surface.

1. Introduction

Some optical devices and sensors, such as dye-sensitized solar cells (DSSC), optical gas sensors, biochemical sensors, are built by a mesoporous layer as its active layer. [1-7] The thickness of the layer may be in a few micrometers, but it is also often in a few tens of nanometer. Small dye molecules are often impregnated into the mesoporous layer as active molecules or sensing molecules. While the dye deposition on the surface can be confirmed by some microscopy techniques, such as scanning electron microscopy (SEM) or atomic force microscopy (AFM), the dye adsorption inside the mesoporous layer is not easy to be confirmed. For that purpose, Surface Plasmon Resonance (SPR) spectroscopy may provide an alternative way due to large penetration depth of surface plasmon wave on the metal/dielectric interface.

SPR is an evanescent wave generated by resonance of an incident *p*-polarized light wave resonances with plasmon oscillation occurred at the interface between thin metal film (ε_m) and dielectric (ε_d). SPR can be generated by various means by using a prism coupler or a grating coupler, waveguide, structure and metal nanoparticle structure. In case of SPR generation by using a prism coupler, the resonance condition is given by

$$\frac{\omega}{c}\sqrt{\varepsilon_p}\sin\theta_o = \frac{\omega}{c}\sqrt{\frac{\varepsilon_m\varepsilon_d}{\varepsilon_m+\varepsilon_d}} \tag{1}$$

where ω is the angular frequency, c is the speed of light, ε_p is the dielectric permittivity of the prism, θ_o is the incident angle of the *p*-polarized incident light, ε_m is the dielectric permittivity of the metal layer and ε_d is the dielectric permittivity of the dielectric layer. While in many cases we found the dielectric layer is not a light absorbing material, in the present work we consider it is a light absorber

IOP Conf. Series: Journal of Physics: Conf. Series **1057** (2018) 012002 doi:10.1088/1742-6596/1057/1/012002

due to the dye molecules in the mesoporous layer. In such case, the dielectric layer becomes an absorbing layer. The eq. (1) is still applicable, but the dielectric permittivity ε_d is now a complex number. In this paper, we report our evaluation on the possibility of SPR spectroscopy for detecting the adsorption of dye molecules inside a mesoporous layer.

2. Methodology

2.1. Experimental

Figure 1. Illustration of the a) adsorbed type, and b) adsorbed and surface deposited type of dye in AZO mesoporous layer, respectively.

In this work, we consider two types of dye adsorption cases, namely the adsorbed type (type I) and the adsorbed and surface deposited type (type II), as illustrated in Figure 1. The SPR generation in those structures was carried out by using a prism coupler in the Kretschmann configuration. In such configuration, in order to have resonance wavelength at visible region, the dielectric permittivity of the substrate and the dielectric layer should be smaller than the permittivity of the prism coupler. Therefore, in this work, we controlled the AZO layer to be thin enough so that the thickness was quite smaller compared to the penetration depth of the SPR wave. In such case, the permittivity of the AZO layer was smaller than the prism coupler permittivity (ε_p= 2.25), although the bulk AZO permittivity is around 3.24.

2.1.1. Samples

Au thin layer with the thickness of 50 nm was deposited onto a glass substrate by sputtering method. The AZO precursor solution was synthesized by a method reported elsewhere by dissolving 0.25 M zinc acetate dehydrate ($(Zn(CH_3COO)_2).2H_2O$) in 2-methoxyethanol with diethanolamine (DEA) as the stabilizing agent and $AlCl_3$ as the Al dopant. [8-9] The solution was stirred for 24 hours. This AZO precursor solution was then spin coated onto an Au thin layer, which was subsequently subjected to heat treatment up to 500°C. The thicknesses of these AZO layers were about a few tens nanometer, which was supposed to be much smaller than the penetration depth of the SPR wave. These AZO layers were then immersed in dye N719 solution in 0.25 hour and an hour.

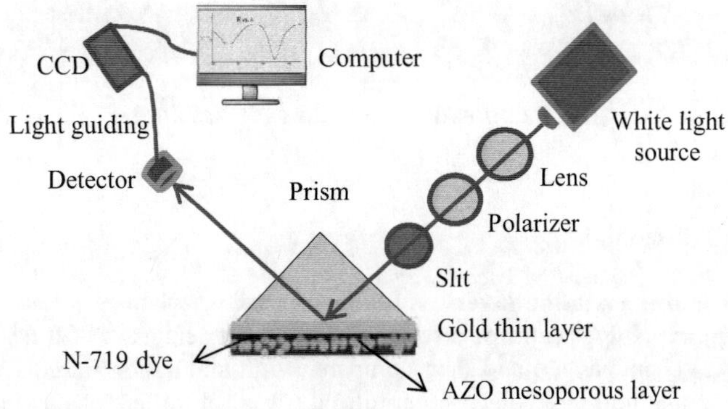

Figure 2. The experimental set-up of SPR spectroscopy used in this work.

IOP Conf. Series: Journal of Physics: Conf. Series **1057** (2018) 012002 doi:10.1088/1742-6596/1057/1/012002

2.1.2. Experimental set-up

The measurement of SPR spectrum was performed by using the *p*-polarized polychromatic light from a tungsten lamp in Kretschmann configuration at several incident angles. The reflected beam from the SPR element was captured by a light guiding and then delivered to the CCD based spectrophotometer, as indicated in Figure 2.

2.2. Simulation

The SPR spectra were simulated by using a simple calculation based on the transfer-matrix method for multilayer structures at oblique incidence. The relationship between the electric field at the *i-th* layer and the nearest adjacent (i+1)-*th* layer is given by [10,11]

$$\begin{bmatrix} E_i^+ \\ E_i^- \end{bmatrix} = \frac{1}{\tau_i} \begin{bmatrix} e^{jk_i l_i} & r_i e^{-jk_i l_i} \\ r_i e^{jk_i l_i} & e^{-jk_i l_i} \end{bmatrix} \begin{bmatrix} E_{i+1}^+ \\ E_{i+1}^- \end{bmatrix} \tag{2}$$

The total reflection coefficient is then given by

$$\Gamma_i = \frac{r_i + \Gamma_{i+1} e^{-2jk_i l_i}}{1 + r_i \Gamma_{i+1} e^{-2jk_i l_i}} \Rightarrow \Gamma_1 = \frac{r_1 + \Gamma_2 e^{-2jk_1 l_1}}{1 + r\Gamma_2 e^{-2jk_1 l_1}}$$

$$\Gamma_i = \frac{E_i^-}{E_i^+} = \frac{\frac{1}{\tau_i}\left(r_i e^{jk_i l_i} E_{i+1}^+ + e^{-jk_i l_i} E_{i+1}^-\right)}{\frac{1}{\tau_i}\left(e^{jk_i l_i} E_{i+1}^+ + r_i e^{-jk_i l_i} E_{i+1}^-\right)} \tag{3}$$

and the reflectance R is given by

$$R = |\Gamma_p|^2 = \left|\prod_{i=1}^{N-1} \Gamma_i\right| \tag{4}$$

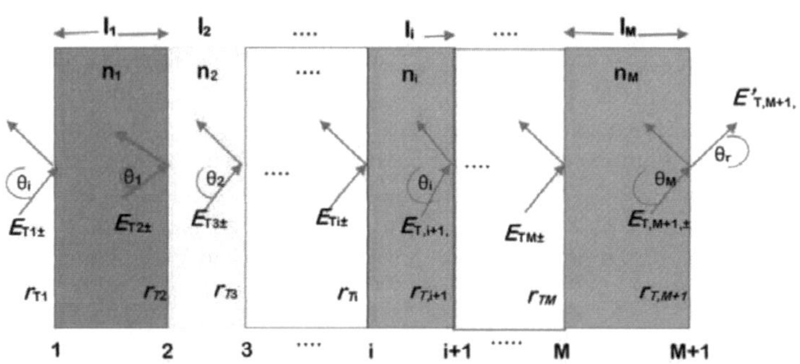

Figure 3. Electric field and other notations in multilayer structure model used in the simulation.

3. Results and discussion

3.1. SPR spectra of the Au thin layer

As the first step, the SPR spectra of an Au layer with a thickness of 50 nm in the prism/Au/air configuration was characterized and then compared with simulation based on the transfer matrix method. In the simulation, the dielectric permittivity was set to be close as in the experiment, i.e. $\varepsilon_{prism}=2.25$, and $\varepsilon_{air}=1$, while ε_{Au} was calculated based on the Drude-Lorentz model. Figure 4 shows that the spectra obtained from the experiment are almost similar as obtained from the simulation. At

IOP Conf. Series: Journal of Physics: Conf. Series **1057** (2018) 012002 doi:10.1088/1742-6596/1057/1/012002

the same incident angle, the SPR dip, which is indicated by the reflectance dip, appears at almost the same wavelength in both experimental and simulation spectra. The dip is blue shifted with increasing incident angle.

3.2. SPR spectra of the Au/AZO layer

Figure 5.(a) shows the SPR spectra of the Au/AZO layer in the prism/glass/Au/AZO/air configuration, which seems different from the Au layer spectra in Figure 4.(a). The dips are shifted to longer wavelength for the same incident angle, which indicates larger permittivity due to the presence of AZO thin layer. However, the AZO layer is quite thin so that the resonance wavelengths are still in the visible region. Figure 5.(a) shows the simulation results with the parameters: ε_{AZO} = 3.2400 and the AZO thickness is about 20 nm. In this simulation, we disregarded the wavelength dependence of the dielectric permittivity of the AZO layer. The dielectric permittivity was considered to be almost constant in the wavelength region considered here. We can see clearly that the dip wavelengths are almost similar in the experimental and the simulation spectra.

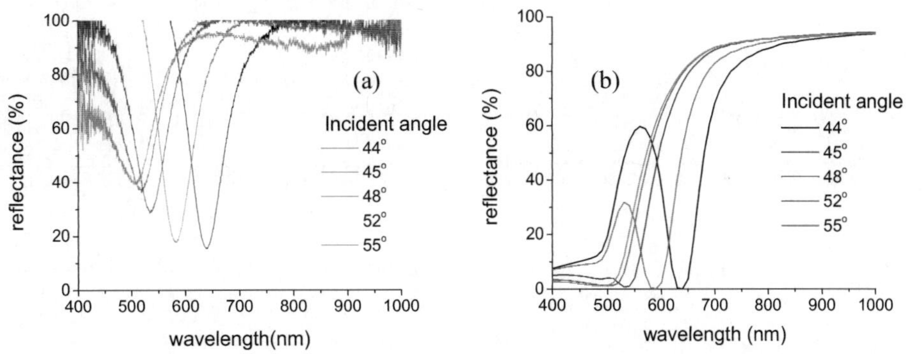

Figure 4. The SPR spectra of an Au layer with a thickness of 50 nm in the prism/glass/Au/air configuration obtained from a) experiment and b) simulation.

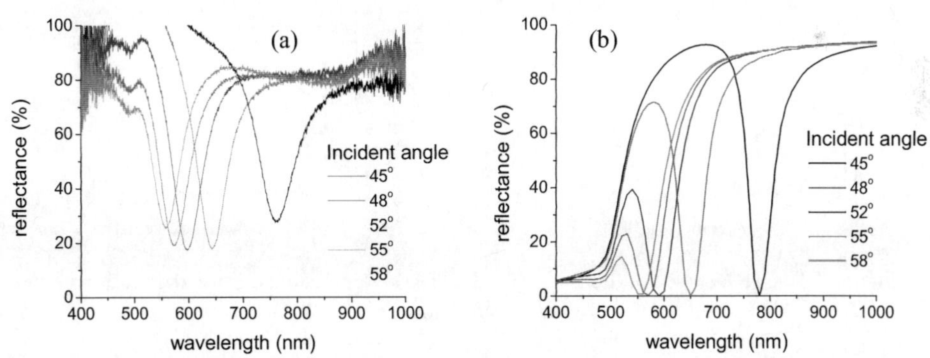

Figure 5. The reflectance of the prism/glass/Au/AZO/air configuration obtained from a) experiment and b) simulation.

3.3. SPR spectra of the Au/AZO+dye layer

After the immersion in N719 dye solution in 0.25-hour, the sample color changed into transparent red color, indicating a dielectric permittivity with a complex number consisting of the real and imaginary values. It is, however, not easy to determine it experimentally. We suppose that in this situation the sample is the type I in Figure 1, which is the porous adsorbed type. Figure 6.(a) shows

IOP Conf. Series: Journal of Physics: Conf. Series **1057** (2018) 012002 doi:10.1088/1742-6596/1057/1/012002

the SPR spectra where the dip is further shifted to a longer wavelength and the dip becomes shallower. This is due to the increase of both real and complex parts of the total permittivity caused by the dye infiltration into the AZO mesoporous layer. The SPR spectra from the simulation seem similar to the experimental results, where in this simulation we used the parameters: $\varepsilon_{AZO+dye}$ = 3.9998+0.08 i and the AZO+dye layer thickness is 20 nm. Here, we considered again that the dielectric permittivity is almost constant in the evaluated wavelength region. The imaginary value indicates the absorption effect from the dye.

3.4. SPR spectra of the Au/AZO+dye/dye layer

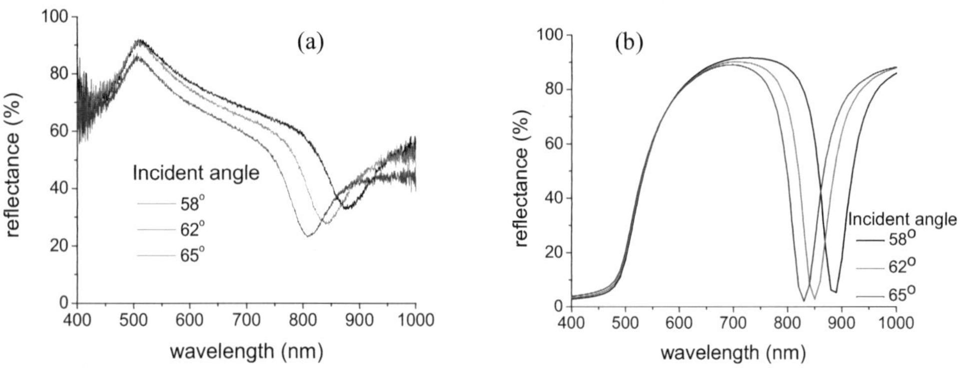

Figure 6. The SPR spectra of Au/AZO+dye layer obtained from a) experiment, and b) simulation.

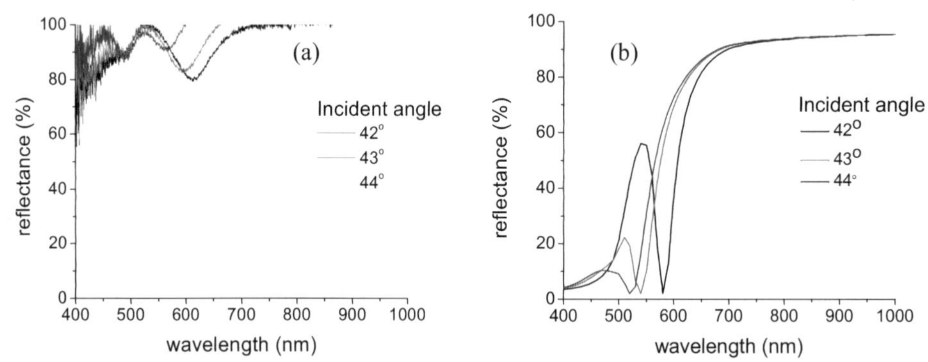

Figure 7. The reflectance of the Au/AZO+dye/dye layers obtained from a) experiment and b) simulation.

Immersion in into the dye solution in a longer time, that is, in an hour, seems to produce the SPR spectral dips appeared in the shorter wavelength region. The dips now appear at around 600 nm for incident angels around 42°-44°. This shifting seems unusual. However, in this case, the sample color visually become darker red color, supposing the dye deposition on the surface of the AZO layer, as illustrated as the type II in Figure 1. The simulation results nevertheless support that layer formation as shown in Figure 7.(b). The simulation shows the appearance of SPR dip around 600 nm obtained by using dielectric permittivity parameters: $\varepsilon_{AZO+dye}$ = 3.9998 + i 0.08 with the AZO+dye thickness is 20 nm and ε_{dye} = 6. 2496 + i 0.1 with the surface deposited dye layer thickness is 30 nm.

Finally, the agreement between the experimental and the simulation results are shown in Figure 8, which is indicated by the points scatter around the theoretical curve lines in the dispersion curve.

The theoretical dispersion curve lines calculated from the eq. (1) are also shown for dielectric constant ε_d = 1 (for prism/Au/air) and for effective dielectric permittivity ε_d = 1.25 (for prism/Au/AZO) as a comparison. The effective dielectric permittivity here is much smaller than the dielectric permittivity of AZO bulk, which indicates that the AZO layer is extremely thin so that the upper portion of SPR wave widely overlaps with the air, as illustrated in Fig. 1.

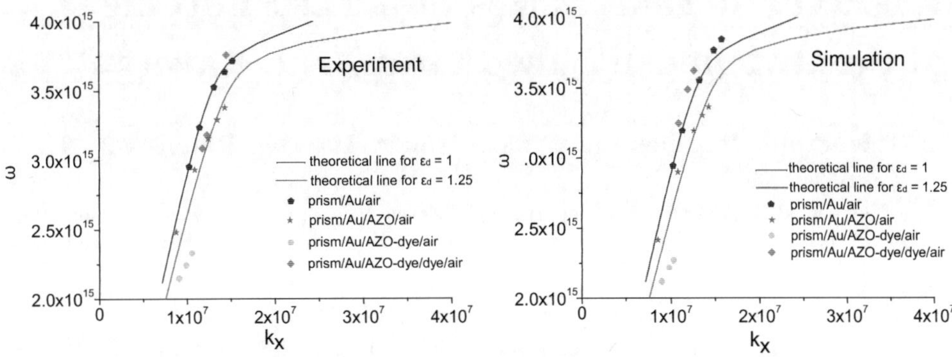

Figure 8. The dispersion curves obtained from a) the experiment and b) simulation results. The theoretical dispersion curve lines calculated from the eq. (1) for ε_d = 1 and 1.25 are also shown for comparison.

4. Conclusion

The present work shows that this SPR spectroscopy can be used effectively as a non-destructive tool for identifying the adsorption of dye molecules inside a mesoporous layer. A combined method of experiment and simulation can distinguish well the two different types of dye adsorption in the mesoporous structure.

5. References

1. Sanzaro S, Mecca E, Mannino G, Bongiorno C, Pelligrino G, Neri F, Malandrino G, Catalano M R, Condorelli G G, Lacobellis R, *et.al.* 2016 *Scientific Report* **6** 39509
2. Wijaya E, Lenaerts C, Maricot S, Hastanin J, Habraken S, Vilcot J P, Boukherroub R, Szunerits S 2011 *Solid States and Material Sciences* 208-224
3. Gusak V, Heineger L P, Graetzel M, Langhammer C, Kasemo B 2012 *Nano Lett.* **12** 2397-2403
4. Patskovsky S, Dallaire A M, Meunier M 2016 *Sensor and Actuator* B **222** 71-77
5. Tumkur T U, Zhu G, Noginov M A 2016 *Optics Express 3921* **24** 4
6. Arsyad W S, Bahar H, Prijamboedi B, Hidayat R 2018 *Ionics* **24** 901-914
7. Pujiarti H, Arsyad W S, Wulandari P, Hidayat R 2014 *AIP Conf. Proc.* **1617** 39-42
8. Handayani Y S, Wulandari P, Hidayat R 2015 *Advanced Material Research* **1112** 251-255
9. Aprilia A, Wulandari P, Suendo V, Herman, Hidayat R, Fujii A, Ozaki M 2013 *Solar Energy Materials & Solar Cells* **111** 181-188
10. Kadlec R, Kroutilova E, Fiala P 2011 *PIERS Online* **7** 633-636
11. Orfanidis S, Electromagnetic Waves and Antennas, 1031, 2008, accessed at http://www.ece.rutgers.edu/~orfanidi/ewa/

Acknowledgment

The authors acknowledge the Program Proyek Pengembangan ITB (III) 2015, JICA – CAS for the support to this research work.

IOP Conf. Series: Journal of Physics: Conf. Series **1057** (2018) 012003 doi:10.1088/1742-6596/1057/1/012003

Transmission Characteristics of a 1D Photonic Crystal Sandwiched by Two Graphene Layers

Hendradi Hardhienata[1], **Angga Ismi Aziz**[1], **Dwi Rahmawati, Husin Alatas**

Theoretical Physics Division, Department of Physics, Bogor Agricultural University, Jl. Meranti, Gedung Wing S, Kampus IPB Darmaga, Bogor 16680, Jawa Barat, INDONESIA

E-mail: hendradi@ipb.ac.id

Abstract. We analyze the transmission characteristics and dispersion profile of a terahertz (THz) electromagnetic wave propagating inside a 1D dielectric photonic crystal sandwiched by two single atomic graphene sheets. We vary the thickness and number of alternating slabs of the dielectric medium as well as changing the electrical conductivities in the graphene layers and study their effects on the optical bandgap. The simulation result shows that attachment of two single atomic graphene sheets into the system will cause a decrease in the transmitted field although changes in the electrical conductivity will not alter the bandgap since the graphene layer is very thin relative to the dielectric slabs. This feature is in agreement with the results obtained from the dispersion relation curve which is unchanged by varying the electrical conductivity.

1. Introduction

The rise of modern technology is undeniably impressive and in recent times its evolution has been unprecedented in speed driven by the desire to produce not only strong but also lightweight material. Rapid progress in this field cannot be separated from the ability of physicists in manipulating at the nanoscale which is supported by mastery of solid state physics and optoelectronics. Among many important materials graphene has recently stood out as a promising candidate in shaping our future technology due to its electrical and optical properties.

Graphene is the name given to a flat monolayer of carbon atoms tightly packed into a two-dimensional (2D) honeycomb lattice, and is a basic building block for graphitic materials of all other dimensionalities. Although graphene exhibits almost uniform absorption within a large wavelength range, a single layer of graphene may absorb as much as 2.3% of the incident light.[1] Since its experimental discovery, graphene has attracted significant attention in recent years due to the richness of its optical and electronic properties. The unique band structure of graphene leads to many important potential applications in nanoelectronics, especially in the fabrication of ultrafast transistors. As an ultrathin two-dimensional semiconductor, graphene also has unique optical properties and many important potential applications, such as optical limiters and polarizers. Graphene has no band gap and very high electron mobility of about 15000 $cm^2V^{-1}s^{-1}$ at a temperature of 300 K and about 60000 $cm^2V^{-1}s^{-1}$ at 4 K, whereas for graphene layer is about 3000−10000 $cm^2V^{-1}s^{-1}$.[2] The special properties of graphene can be

[1] These authors contribute equally

IOP Conf. Series: Journal of Physics: Conf. Series **1057** (2018) 012003 doi:10.1088/1742-6596/1057/1/012003

analyzed to determine the optical conductivity of graphene as a function of energy. Without interference, the optical conductivity of graphene has a universal value or in the range of infrared energy to visible light.[3] Lately, a lot of research examines how the band gap of graphene can be opened and controlled to obtain materials in accordance with the application of modern electronic devices.

One of the open challenges is to increase the light absorption in graphene. Several mechanisms have been proposed to enhance the absorption of graphene, including putting graphene inside an optical cavity, exploiting the physics of quantum dots on graphene, using a uniaxial strain, a graphene-based grating, and surface plasmons.[4] Recently it has also been shown that inserting graphene slabs into a 1D photonic crystal can increase field absorption by a factor of four.[5] Indeed, photonic crystals has opened a new door for advanced photonic component due to its amazing properties. Photonic crystals can be found in nature and can be created artificially and posseses the remarkable feature of a forbidden band gap where light propagation is prohibited. Photonic crystals are typically classified as one, two, and three dimensional depending on the periodic layers arranged within the structure. When a defect layer is introduced in a 1D photonic crystal, the passband that occurs inside the transmission bandgap can be controlled by changing the refractive index of the defect layer.[6] However, development of optical integrated circuits has certain problems such as unavailability of such materials which can be engineered for optimal photon trapping and confinement without losses at the sharp corners of the waveguides. In this context, photonic band structures or photonic crystals are the promising one to overcome such difficulties and leading continued progress of advance optical integrated circuits. Intensive research on photonic crystals has been attempted after the pioneer works of Yablonovitch and John in 1980s.[7] This study takes understanding of the electromagnetic waves behavior in a simple photonic crystal (periodic medium bilayer). Simple photonic crystal made of two alternate transparent medium with a different refractive indices. Crystal engineering technology development has enabled the making of medium periodically with periodicity well controlled and the thickness of the layer is only a few times thickness of an atom (e.g. molecular epitaxy technique).

In this work, the study of multi-layer graphene is limited to the transmittance output waveform and the analysis of the dispersion relations. The transmittance was studied using the transfer matrix methods, which have been proven to be effective in calculating the electromagnetics field propagation in dielectric and become the flexible methods for modify the dielectric layer as well as the sandwiched by two graphene layers. Electromagnetic waves are propagated in a one-dimensional photonic crystals can cause thin transmittance areas/bandpass inside the forbidden band. The position of the passband inside the photonic bandgap can also be determined from the dispersion relation.[8] The trasmission properties of one dimensional graphene photonc crystals were investigated by using transfer matrix method and it was found that the structure has new type of the photonic bandgap in the terahertz (THz) region and is insensitive to the polarization. The reason we perform our simulation in the terahertz spectra is that in this region the properties of the photonic band gap depend on the optical parameters of the graphene sheets.[9] These properties of 1D photonic crystals have potential applications in tunable terahertz multiband narrow filters.[10] In the following section we describe the basic theory in obtaining the transmission profile and dispersion relation via the transfer matrix method (TMM).

2. Transfer Matrix Method

The transfer matrix method is developed based on the electromagnetic boundary conditions that Maxwells equations required. Compared with the traditional transfer matrix method, the new method can be used to calculate the optical absorption of any layer at any position in stratified media.[11] The transfer matrix method is a method used in optics and acoustics to analyze the propagation of electromagnetic or acoustic waves through a stratified (layered) medium. The transfer-matrix method is based on the fact that according to Maxwell's equations, there are simple continuity conditions for the electric field across boundaries from one medium to the next. If the field is known at the beginning of a layer, the field at the end of the layer can be derived from a simple matrix operation. A stack of layers can then be represented as a system matrix, which is the product of the individual layer matrices. The final step of the method involves converting the system matrix back into reflection and transmission coefficients. Figure 1 shows our model where the single atomic graphene layer structure is sandwiched near both ends of the 1D PC.

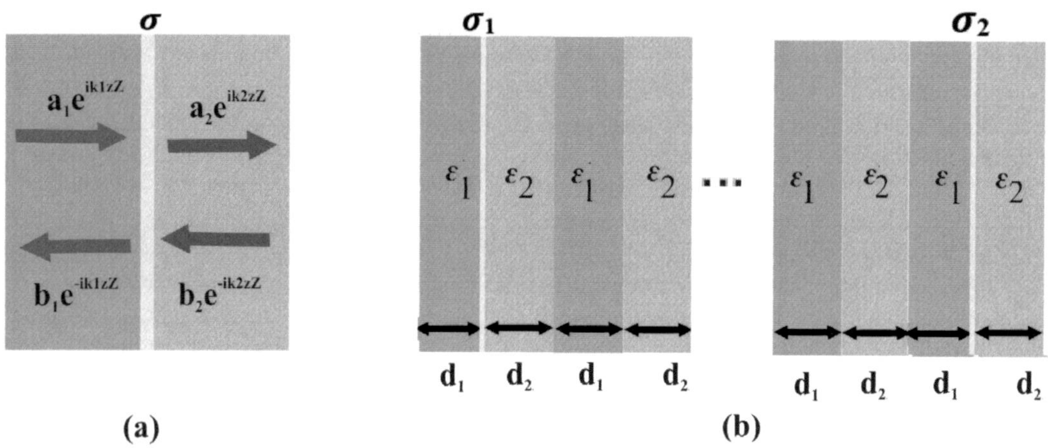

Figure 1: 1D PC structure assumed in the model. (a) The transfer matrix unit cell containing graphene layer (yellow layer) sandwiched between two dielectric materials (blue and purple slabs) with dielectric constants ε_1 and ε_2. Here the arrows to the right and left indicate the transmitted and reflected light respectively (b). Arrangement of the 1D PC and the position of the two graphene layers. The background is assumed to be air ($\varepsilon_0 = 1$).

In this study only electromagnetic waves with polarization p or TM mode (Transverse Magnetic) is considered where the magnetic field is polarized in the y direction and propagatong in the z-axis, therefore the field can be written as follows:

$$H_{1y} = (a_1 e^{ik_{1z^z}} + b_1 e^{-ik1z^z}) e^{ik_{1x^x}} z < 0, \tag{1}$$

$$H_{2y} = (a_2 e^{ik_{2z^z}} + b_2 e^{-ik2z^z}) e^{ik_{2x^x}} z > 0, \tag{2}$$

a_1 and b_1 (i=1,2) is the field coefficient,k_{ix} (k_{iz}) is the component x (z) of the wave vector $k_{ix}=\sqrt{\varepsilon_i}\frac{\omega}{c}$ which is the angular frequency and c the speed of light in a vacuum. The electric field and magnetic field at the interface of a dielectric can be connected to the boundary conditions as follows:

$$nx(E_2 - E_1)|_{z=0} = 0, \tag{3}$$

$$nx(H_2 - H_1)|_{z=0} = J, \tag{4}$$

Where n is the normal unit of the plane and the current density of the graphene layer. From the relationship of equations (3), (4) and equations (1), (2) using boundary conditions in obtain the following matrix equations:

$$\frac{k_{iz}}{\varepsilon_1}(a_1 - b_1) - \frac{k_{2z}}{\varepsilon_2}(a_2 - b_2) = 0, \tag{5}$$

IOP Conf. Series: Journal of Physics: Conf. Series **1057** (2018) 012003 doi:10.1088/1742-6596/1057/1/012003

$$(a_1 - b_1) - (a_2 - b_2) = J_x, \tag{6}$$

With J given from Ohm's law so the equation J_x Where ε_0 is the permittivity in vacuum. From the equations (5) and (6) the field coefficients a_1 and b_1 can be related to the field coefficients a_2 and b_2 through the transmission matrix 2x2 as follows:

$$\begin{pmatrix} a_1 \\ b_1 \end{pmatrix} = D_{1\longrightarrow 2} \begin{pmatrix} a_2 \\ b_2 \end{pmatrix} \tag{7}$$

where $D_{1\longrightarrow 2}$ is:

$$D_{1\longrightarrow 2} = \frac{1}{2} \begin{pmatrix} 1 + \eta_p + \xi_p & 1 - \eta_p - \xi_p \\ 1 - \eta_p + \xi_p & 1 + \eta_p - \xi_p \end{pmatrix} \tag{8}$$

with parameter η_p and ξ_p is:

$$\eta_p = \frac{\varepsilon_1 k_{2z}}{\varepsilon_2 l_{1z}} \tag{9}$$

$$\xi_p = \frac{\sigma k_{2z}}{\varepsilon_0 \varepsilon_2 \omega} \tag{10}$$

and the propagation matrix is :

$$P(\Delta z) = \begin{pmatrix} e^{-ik_z \Delta z} & 0 \\ 0 & e^{ik_z \Delta z} \end{pmatrix} \tag{11}$$

Thus the transfer matrix is the multiplication of the transmissi matix with the propagation matrix on the homogeneous dielectric material. With the note of the field coefficient a_1 and b_1 to the left of the grahene layer and the field coefficient a_{N+1} and b_{N+1} to the right of the gaphene layer so that the two field coefficients are related to the transfer matrix (β) 2x2 is:

$$\begin{pmatrix} a_1 \\ b_1 \end{pmatrix} = \beta \begin{pmatrix} a_{N+1} \\ b_{N+1} \end{pmatrix} \tag{12}$$

with

$$\beta = D_{1\longrightarrow 2} P(d_{1,2}) D_{2\longrightarrow 3} P(d_{2,3}) ... P(d_{N-1,N}) D_{N,N+1} \tag{13}$$

Note that since the structure is periodic, then $d_{1,2} = d_2$ and $d_{2,3} = d_1$. Here β is a 2 x 2 matrix:

$$\beta = \begin{pmatrix} \beta_{11} & \beta_{12} \\ \beta_{21} & \beta_{22} \end{pmatrix} \tag{14}$$

thus,

$$\begin{pmatrix} a_1 \\ b_1 \end{pmatrix} = \begin{pmatrix} \beta_{11} & \beta_{12} \\ \beta_{21} & \beta_{22} \end{pmatrix} \begin{pmatrix} a_{N+1'} \\ b_{N+1'} \end{pmatrix} \tag{15}$$

if both of segments are divided by a'_{N+1} the matrix is:

$$\begin{pmatrix} \dfrac{a_1}{a_{N+1'}} \\ \dfrac{b_1}{a_{N+1'}} \end{pmatrix} = \begin{pmatrix} \beta_{11} & \beta_{12} \\ \beta_{21} & \beta_{22} \end{pmatrix} \begin{pmatrix} \dfrac{a_{N+1'}}{a_{N+1'}} \\ \dfrac{b_{N+1'}}{b_{N+1'}} \end{pmatrix} = \begin{pmatrix} \beta_{11} & \beta_{12} \\ \beta_{21} & \beta_{22} \end{pmatrix} \begin{pmatrix} 1 \\ 0 \end{pmatrix} \begin{pmatrix} \beta_{11} & \beta_{12} \\ \beta_{21} & \beta_{22} \end{pmatrix} = \begin{pmatrix} \beta_{11} \\ \beta_{21} \end{pmatrix} \tag{16}$$

The transmittance is given by:

$$t = \left(\frac{1}{\beta_{11}} \right) \tag{17}$$

IOP Conf. Series: Journal of Physics: Conf. Series **1057** (2018) 012003 doi:10.1088/1742-6596/1057/1/012003

3. Dispersion Relation for Simple 1D Photonic Crystals

In general, wave propagation in periodic media can be described in terms of Bloch waves. For a determination of the dispersion of a periodic crystal, it is necessary only to integrate the wave-field through a periodic media. Periodicity requires that the Bloch wave must obey the translation symmetry $z = z + I\Lambda$, where I is an integer. Here we follow the derivation of the dispersion relation in accordance with Yariv and Yeh.[12] According to the Bloch theory, the electric field vector which propagate on a line to the z-axis (component ky = 0) in the form of periodic photonic crystals:

$$E = E_k(z)e^{iK.Z} \tag{18}$$

where the K subscript here indicates that the function of E_k depend on Bloch wave vector K. The problem that needs to be solved is how to determine K as a function of ω, or otherwise. The relation between the variables ω and K is called the dispersion relation. Mathematically it can be written as:

$$\omega = \omega(k) \tag{19}$$

In matrix notation the equation would be:

$$\begin{pmatrix} A_{N^{(1)}} \\ B_{N^{(1)}} \end{pmatrix} = e^{ikz} \begin{pmatrix} A_{N-1^{(1)}} \\ B_{N-1^{(1)}} \end{pmatrix} \tag{20}$$

whereas A_N is the amplitude of the electric field that is passed on to the K N-layers, while B_N is reflected electric field amplitude. Eq. (20) can be combined with eq. (18) to produce:

$$\begin{pmatrix} A & B \\ C & D \end{pmatrix} \begin{pmatrix} A_{N^{(1)}} \\ B_{N^{(1)}} \end{pmatrix} = e^{ikz} \begin{pmatrix} A_{N-1^{(1)}} \\ B_{N-1^{(1)}} \end{pmatrix} \tag{21}$$

A phase factor $e^{iK\Lambda}$ and the matrix components (A_N, B_N), the equation (21) satisfies the seculer equation:

$$\begin{pmatrix} A - e^i k\lambda & B \\ C & D - e^i k\lambda \end{pmatrix} \tag{22}$$

The secular equation can be obtained easily from linear algebra because $Mx = \lambda$x (with λ=$e^{iK\Lambda}$) is an Eigen equation, then $(M - \lambda\text{I})$x $= 0$ or in order words det $(M - \lambda\text{I})$ should be zero. Then the the seculer equation takes the form[12]:

$$e^{ik\Lambda} = \frac{1}{2}(A + D) \pm \left[\left(\frac{1}{2}(A + D)\right)^2 - 1\right]^{\frac{1}{2}} \tag{23}$$

Equation (23) is the dispersion relation between ω, ky and K for the Bloch wave. The equation consists of real and imaginary components. By taking the real components the following is obtained:

$$K(k_y, \omega) = \frac{1}{\lambda} cos^1 \left[\frac{1}{2}(A + D)\right] \tag{24}$$

In areas where $\left[\frac{1}{2}(A + D)\right] < 1$ the K value is real while for $\left[\frac{1}{2}(A + D)\right] > 1$ the K value is complex and for the latter case Bloch wave disappears hence no photon with momentum $\hbar k$ can propagate and we observe a forbidden band or band gap. The edges of forbidden band occurs at $\left[\frac{1}{2}(A + D)\right] = 1$. Changing the refractive index n_1 and n_2 as well as the distance layer, a and b

IOP Conf. Series: Journal of Physics: Conf. Series **1057** (2018) 012003 doi:10.1088/1742-6596/1057/1/012003

will change the value of transfer matrix elements A and D therefore changing the width of the forbidden band. Applying the notation convention in eq. (14) the dispersion relation takes the form:

$$CosK\Lambda = \left(\frac{\beta_{11} + \beta_{22}}{2}\right) \tag{25}$$

Eq. (25) will be used in generating the dispersion relation in the simulation.

4. Results and Discussions

In the following section we decribe the tarnsmission profile for variation of the dielectric medium thickness, number of layers, and electrical conductivity.

4.1. Variation of dielectric slab thickness

In the simulation we set the 1st and 2nd dielectric medium refractive index to 1 and 3 respectively and set $\sigma = 0$. Here d_1 is the thickness of the 1st dielectric medium, d_2 is the thickness of the 2nd dielectric medium where their dimension is in accordance with the condition $n_i d_i \approx \lambda_0$ where λ_0 is the Bragg wavelength. Here, N is the number of alternating layers (double layers). Assuming no graphene layer, variation of N and dielectric thickness results in changes of the bandgap width as presented in Fig. 4. This result is not new and is indeed well-known, but here we also plot the dispersion relation profile (green curve) to see how the bandgap changes accordingly.

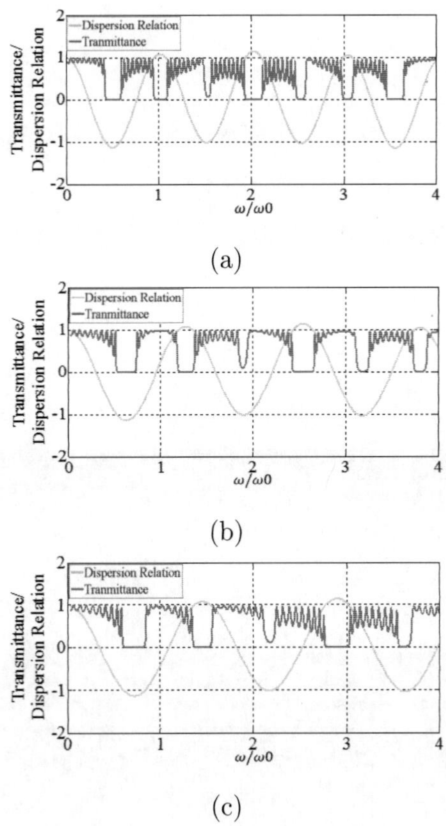

Fig.2. Variation of dielectric slab thickness on the transision profle (a) d_1=0.4 mm, d_2=0.6 mm, N=30, $\sigma = 0$, (b) d_1=0.6 mm, d_2=0.6 mm, N=30, $\sigma = 0$, (c) d_1=0.6 mm, d_2=0.8 mm, N=30, $\sigma = 0$

IOP Conf. Series: Journal of Physics: Conf. Series **1057** (2018) 012003 doi:10.1088/1742-6596/1057/1/012003

From figure 4(a), 4(b), and 4(c), it can be inferred that assuming no graphene layers, the effect of adding more dielectric layers will increases the number of bandgaps for a similar bandwidth range. Applying the dispersion relation in eq. (25) shows that increasing the number of layers will increase the wave number of the cosine function where the region near the cosine peaks exceed ± 1 thus resulting in a complex value to K and in this frequency region ($\cos K\Lambda > \pm 1$) the photon cannot propagate hence we observe an optical bandgap.

4.2. Variation of number of dielectric layers

Keeping the dielectric medium refractive index to have a value of $n_1 = 1$ and $n_2 = 3$ respectively and setting $\sigma = 0$ we vary the number of dielectric layers and observe the field transmission profile. The simulation result is depicted in Fig. 5.

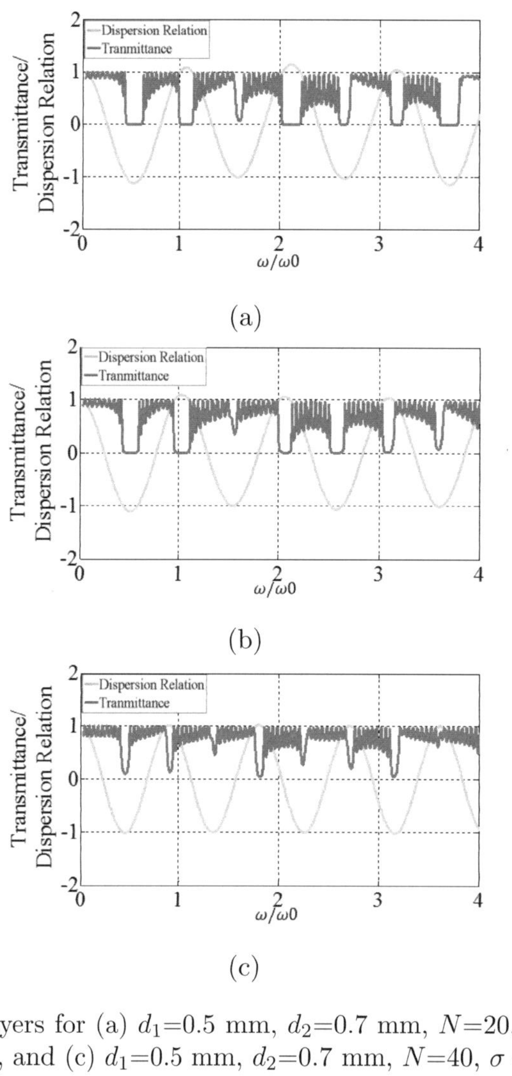

Fig. 3. Variation of N Layers for (a) d_1=0.5 mm, d_2=0.7 mm, N=20, $\sigma = 0$, (b) d_1=0.5 mm, d_2=0.7 mm, N=30, $\sigma = 0$, and (c) d_1=0.5 mm, d_2=0.7 mm, N=40, $\sigma = 0$

As expected increasing the number of dielectric alternating layers will sharpen the bandgaps but has the same bandwidth and increase the number of ripples. Again this result is already well-known, and this can directly be inferred from the dispersion relation curve which stays unchanged.

4.3. Variation of graphene layer conductivity

In the following subsection we keep the 1st and 2^{ndt} dielectric medium refractive index 1 and 3 respectively and set the number of dielectric layers $N = 10$. The thickness of the 1st and 2nd dielectric medium is choosen to be 0.5 mm and 0.8 mm respectively.

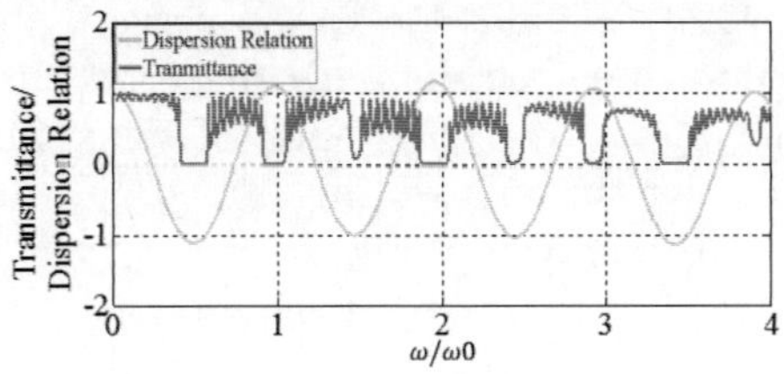

(a)

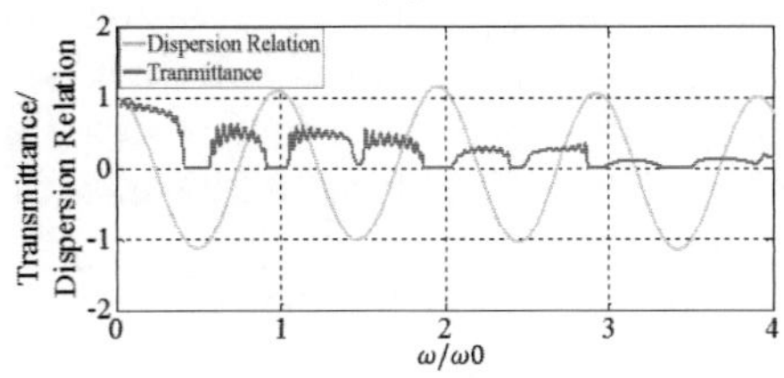

(b)

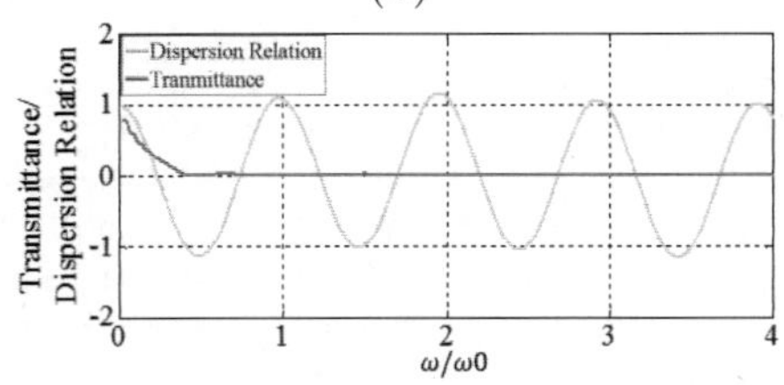

(c)

Fig. 4. Variation of Graphene layer conductivity for (a) $\sigma_1 = 10$, $\sigma_2 = 10$ (b) $\sigma_1 = 100$, $\sigma_2 = 100$ and (c) $\sigma_1 = 1000$, $\sigma_2 = 1000$

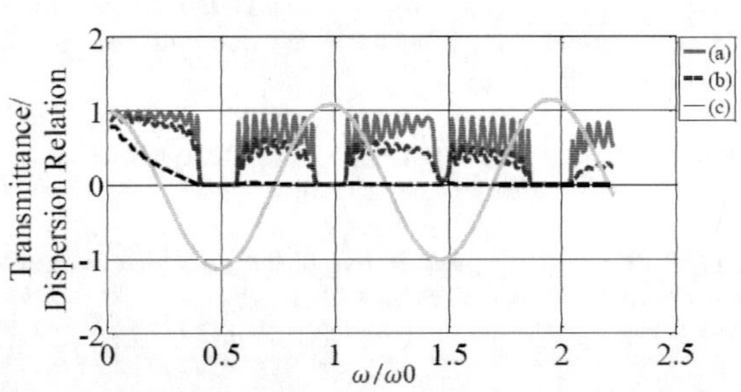

Fig. 5. Graphene layer conductivity variations on the trasmission profile shown here by combining Fig. 4 (a)-(c). Red, blue, and black lines corresponds to Fig. 4 (a), 4(b), and 4(c) respectively whereas the green line corresponds to the dispersion curve which is independent of the graphene conductivity. It is clear that increasing the imaginary part of the graphene conductivity results in higher attenuation due to higher absorption but the bandgap width is unaltered

From figure 4(a), 4(b), and 4(c) by using the variation of the graphene layers conductivity, changing the imaginary part of the graphene electrical conductivity changes the transmittance peak dramatically. This effect can be explained due to the absorption of the field by the graphene slabs along the vertical axis since it was assumed that the thickness of the graphene layer along the propagation direction can be neglected. It is interesting to infer from the dispersion relation curve that although the field is attenuated due to absorption, the bandgap width does not change. Further investigation into the dispersion formula reveal that the effect of changing the electrical conductivity does not affect eq. (25). Only if we attribute certain length for the graphene slab the dispersion relation formula will be altered and thus also the band dap width will change. This feature can neatly be seen in Fig. 5.

5. Conclusion

We have shown that attachment of two very thin graphene layers in a one dimensional photonic crystal alter the transmission profile significantly. Increasing the graphene layer imaginary electrical conductivity value will result in a decrease of the transmission peak although the photonic band gap width remains unaltered. These features can be well explained by the dispersion relation formula which is non sensitve upon changes in the graphene electrical conductivity if the layer thickness is very thin.

Acknowledgements: This research was funded by the "Penelitian Unggulan Perguruan Tinggi 2017 No: 011/SP2H/LT/DRPM/IV/2017". H.H. would like to thank Kurt Hingerl for valuable disscussions.

6. References

[1] Ferreira A., Peres N. M. R., Ribeiro R. M. And Stauber T., Phys. Rev. B, 85 (2012)115438.
[2] Geim A.K, Novoselov K.S. 2007. The rise of graphene. Nature Materials.Vol.6.
[3] Hayt WH jr, Buck JA.2006.Elektromagnetika. Harmein I, penerjemah: hardani HW, editor. Jakarta (ID): Penerbit Erlangga. Terjemahan dari: Engineering Electromagnetics. Ed ke-7.
[4] Joannopoulos J. D, Meade R. D, and Winn J. N, Photonic Crystals Molding the Flow of Light (Princeton University Press, Princeton, (1995).
[5] Leon, S.J. 1986. Linear Algebra with Applications. Macmillan Publishing Company: New York.
[6] Alatas H., Mayditia H., Hardhienata H., A.A. Iskandar., M. O. Tjia.Single-Frequency Refractive Index Sensor Based on a Finite One-Dimensional Photonic Crystals with Two Defects. Japanese Journal of Applied Physics, Vol. 45, No. 8B, 2006, pp. 67546758.
[7] Xin-Hua Deng, Jiang-Tao Liu, Ji-Ren Yuan, Qing-Hua Liao, Nian-Hua Liu. 2015. A new transfer matrix method to calculate the optical absorption of graphene at any position in stratified media. IOP Science.doi: 10.1209/2095-5075/109/27002.
[8] Mayditia H., Hardhienata H., Alatas H. 2015. Analysis of One Dimensional Photonic Crystals with Defect and Asymmetric Edges for Sensor Application. International Conference on Instrumentation, Communication and Information Technology (ICICI). Bandung.
[9] Madani A, Entezar S. R., Characteristics of band structure and surface plasmons supported by a one-dimensional graphene-dielectric photonic crystals. Physica B: Condensed Matter, Vol. 431, 1-5 (2013).
[10] Hajian H., Soltani A., kalafi M.,Characteristics of band structure and surface plasmons supported by a one-dimensional graphene-dielectric photonic crystals. Opt. Commun. 292, 149-157(2013).
[11] Zhan T, Shi X, Dai Y, Liu X, Zi J. 2012. Transfer matrix method for optics in graphene layers. Journal of Physics. 25(21): 1-8.doi: 10.1088/09538984/25/21/215301.
[12] Yariv A, Yeh P. 1984. Optical Waves in Crystal: Propagation and Control of Laser Radiation.Canada (US): John Willey, Inc.

Ab-Initio Computations of Electronic Structures of Methyl-ammonium Lead Bromide/Iodide Perovskites as Wide Bandgap Active Materials in Solar Cells

E D Indari[1], T D K Wungu[2], and R Hidayat[1]

[1] Physics of Photonics and Magnetism Research Division, Physics Program Study, Faculty of Mathematics and Natural Sciences, Bandung Institute of Technology, Jl. Ganesha 10, 40132, Bandung – INDONESIA

[2] Nuclear Physics and Biophysics Research Division, Physics Program Study, Faculty of Mathematics and Natural Sciences, Bandung Institute of Technology, Jl. Ganesha 10, 40132, Bandung – INDONESIA

e-mail: rahmat@fi.itb.ac.id

Abstract. Studies on the electronic structure of organo-lead halide perovskites have been much investigated in the last few years, as solar cells based on this novel materials exhibit high power conversion efficiency. In order to improve the chemical stability and vary the bandgap, for developing a solar cell with bulk heterojunction, the modification of this perovskite with hetero halide ions structure has also been investigated. We report here the electronic structure of Methylammonium lead bromide/iodide or $MAPb(Br_xI_{3-x})_3$ at different contents of bromide ions investigated by an ab-initio computation based on Density Functional Theory (DFT) using Quantum Espresso. The calculated results show the increment of bands gap with increasing bromide ion concentration. The obtained Densities of States show that the addition of bromide ions plays an important role in constructing their valence band, where it is predominated by the electrons at the outermost orbital of these halide (Br^- and I^-) ions. On the other hand, the conduction band is still predominated by the 6p orbitals of the Pb^{2+} ions, in agreement with some reports elsewhere. The organic molecule orbitals, however, seems to give insignificant influence to those valence and conduction bands.

1. Introduction

Development in perovskite solar cells has shown a rapid progress within the last few years after the reports by Snaith and Gratzel groups. [1,2] Inspired by more than 10 % efficiencies performed by the single layer architecture, some researchers are developing perovskite solar cell with tandem and heterostructure layer structure, by combining low and wide bandgap active materials, in order to obtain even higher power conversion efficiency [3,4]. One of the most promising wide bandgap active materials is methylammonium lead bromide/iodide ($CH_3NH_3Pb(Br_xI_{3-x})_3$), or shortly $MAPb(Br_xI_{3-x})_3$. To the best of authors' knowledge, the highest achieved efficiency was just still about 13.7% for this perovskite solar cell [5]. However, it has been also reported that perovskite with such hetero halide ions shows better chemical stability. However, because the most effective method for preparing these materials is by the wet chemical process, there is still remaining a question on the produced crystal structure of this $MAPb(Br_xI_{3-x})_3$. There are two possible products where the first possibility produces a blend of

IOP Conf. Series: Journal of Physics: Conf. Series **1057** (2018) 012004 doi:10.1088/1742-6596/1057/1/012004

nanocrystals of $MAPbBr_3$ and $MAPbI_3$ at a certain ratio, producing a bulk heterojunction layer. The second possibility is nanocrystals of $MAPb(Br_xI_{3-x})_3$ which are formed by interstitial replacements of I anions by Br anions. As we are more concern on the second possible structure, it is necessary to know whether such structure may also exhibit an electronic structure with semiconductor characteristics, that is composed of a conduction band and valence band separated by an energy gap. We may then suppose that the energy gap will be dependent on the ratio of the number of Br anions with respect to the numbers of I anion inside a unit cell. We hereby report our ab-initio computation results based on the density functional theory (DFT) method for this perovskite by varying the content ratio of Br anions.

2. Computational Details

The computations were performed by using Quantum Espresso software package which are built based on the DFT method. Generalized Gradients Approximations (GGA) and Ultrasoft Pseudopotentials (US PPs) were applied. Scalar relativistic exchange-correlation functionals constructed by Perdew-Burke-Ernzerhof (PBE) were applied to each constituent atom. Hence, the valence orbitals comprised of 2s and 2p of C, 1s of H, 2s, and 2p of N, 5d, 6s, and 6p of Pb, 4s and 4p Br, 5s and 5p of I are included. There are four different crystals of $MAPb(Br_xI_{3-x})_3$ that have been investigated, namely for x=0, 0.33, 0.67 and 1. The details are presented in Table 1.

The kinetic energy cut-offs for each wave-functions were limited to 40 Ry to attain precise yet plausible computation cost. For the same reason, the Brillouin zones were divided by Monckhorst Pack k-mesh with 4×4×3 grid in the case of $MAPbI_3$ (or shortly MAPI). More details of this structure can be found in our previous report [6]. Other current investigated structures (see table 1) were reported to have cubic structures [7]. Thus, a homogenous Monckhorst Pack k-mesh grid of 3×3×3 was chosen. The chosen grids were applied for both structure optimizations and electronic structure calculations.

Due to the presence of Pb atoms in the crystal structures, a smearing type of Methfessel-Paxton was employed [7]. The declared lattice constants were extracted from experimental results reported by other groups [8-10]. A pre-optimization of organic molecules' position was performed by applying DFT method with B3LYP (Becke, three parameters, Lee-Yang-Parr) exchange-correlation functional [11]. The unit cell of each crystal structure is depicted in Figure 1 and their initial lattice parameters are shown in Table 1.

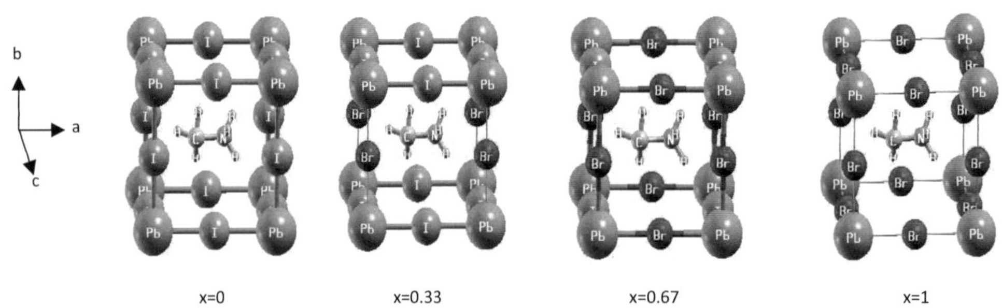

Figure 1. Crystal structures of $MAPb(Br_xI_{3-x})_3$ crystals investigated here. The x-values represent the ratio content of Br/I atom in the $MAPb(Br_xI_{3-x})_3$ crystals.

3. Results and Discussion

The unit cell volume after relaxation and the total energy of each system is depicted in Table 2. From Table 2, one can see that the unit cell volume decreases by increasing the ratio of Br anion. The least total energy value was performed in the MAPI calculation. Although the choice of using MAPI as active materials for perovskite solar application is still facing the chemical stability issue, this material shows the best crystal structure stability compared to the wider band gap methyl-ammonium lead bromide/iodide alternatives [12-13].

Table 1. Crystal structures and lattice parameters of each computed unit crystal.

	Crystal Structure	Lattice Parameter (Å)
MAPI [6]	Pseudocubic	a=6.2621, c =6.3275
$MAPI_2Br$(x=0.33) [9]	Cubic	6.18
$MAPIBr_2$(x=0.67) [9]	Cubic	6.07
$MAPBr_3$ [10]	Cubic	5.90

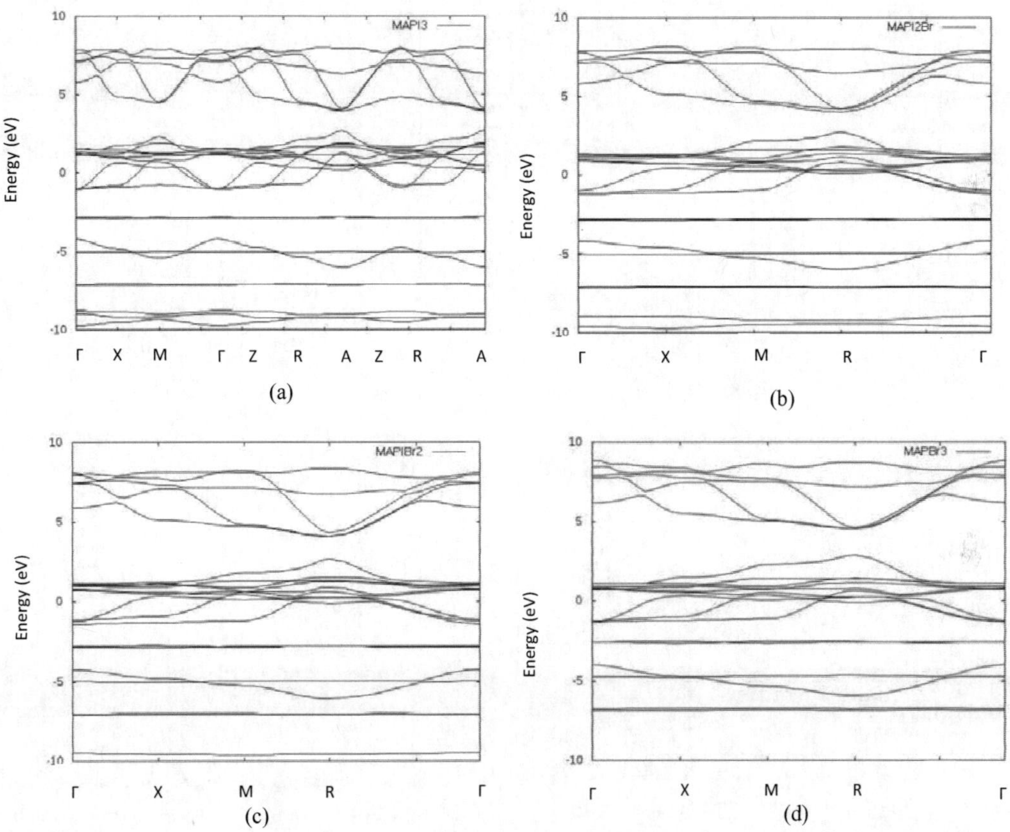

Figure 2. Electronic structure of (a) MAPI, (b) $MAPI_2Br$, (c) $MAPIBr_2$, and (d) $MAPBr_3$.

Table 2 shows the bandgaps values obtained in this work in comparison to the result by Noh *et al.*, where the energy gap was formulated as $E_g(x) = 1.57 + 0.39x + 0.33x^2$, where x is the ratio of Br anions compared to the I anions. Lower values obtained here are due to the application of generalized gradient approximation (GGA) as the exchange-correlation functional [8]. However, one can still see that the obtained bandgap increases as the Br anion numbers increases. As reported by Hao *et al.* in the case of $CH_3NH_3Sn_{3-x}I_x$, the increase of bandgap was found also as the consequence of Br anions addition in our results can be explained by the higher total energy computed [14].

Table 2. Computational results.

	Volume ($Å^3$)	Total Energy (eV)	Obtained Bandgap (eV)	Theoretical Bandgap (eV)
MAPI	248.126	-363.087	1.232	1.570
$MAPI_2Br$(x=0.33)	236.029	-338.078	1.266	1.734
$MAPIBr_2$(x=0.67)	223.648	-313.058	1.428	1.980
$MAPBr_3$	205.379	-288.049	1.636	2.290

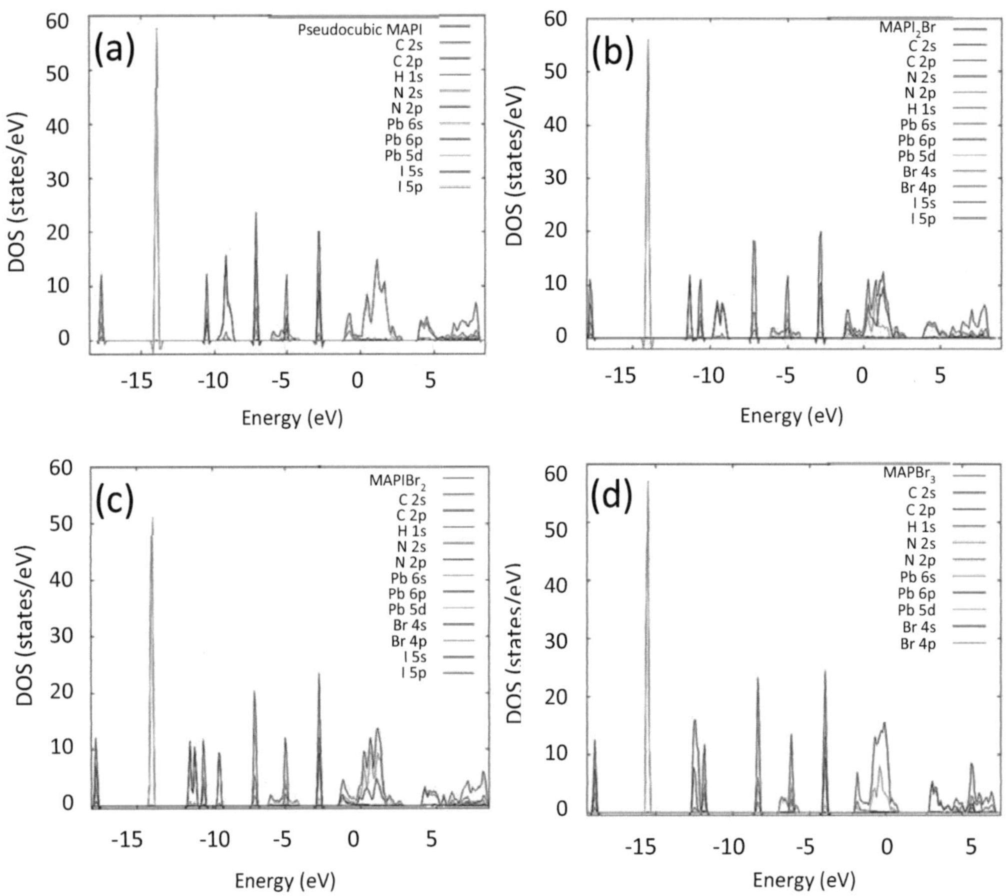

Figure 3. PDOS of (a) MAPI, (b) $MAPI_2Br$, (c) $MAPIBr_2$, and (d) $MAPBr_3$

As shown in Figure 2, MAPI configuration shows a direct bandgap at the A high symmetry point while the rests are at the R high symmetry point. Both the tops of valence and conduction bands of each configuration are quite broad, indicating that the electrons are non-localized. The broadness of valence and conduction bands have been reported as the origin of long exciton diffusion in these perovskites.

Accordingly, the combination of bromide/iodide can increase the exciton transport distance in comparison to that of MAPI due to the broader bands. Besides, the bottoms of conduction bands fluctuate significantly, with MAPI as the most fluctuate. This fact explains the experimental observation of fast photoelectrons transport, where MAPI exhibit the fastest transport speed [15].

The projected densities of states (PDOS) of each configuration are presented in

Figure 3. The valence is predominantly constructed by the outermost orbitals of both halide ions (4p Br^- and 5p I^-). On the other hand, the conduction band is predominantly constructed by the 6p orbital of the Pb^{2+} ions, as reported elsewhere. The contribution of organic molecule orbitals in both valence and conduction bands are very minor, as also shown in our previous report [6].

4. Conclusions

In summary, our computational results show that the bandgap energy of methylammonium lead bromide/iodide ($MAPb(Br_xI_{3-x})_3$) enlarge with the increase of Br^- concentration as the results of the increase in total energy. However, the energy gap does not increase linearly with the content of Br^-. All configurations exhibit non-localized electrons in both valence and conduction bands. Meanwhile, the PDOS graphs show that the outermost orbitals of halide (Br^- and I^-) ions and Pb^{2+} 6p predominantly contribute to the valence and conduction bands, respectively. The organic molecule orbitals contributions to those bands are very minor.

Acknowledgments

Authors acknowledge Program Hibah Kerjasama Luar Negeri 2015/2016 contract no. 077/SP2H/LT/DRPM /II/2016 for the financial support.

References

[1] J. Burschka, et al. *Nature*, 2013, 499, 316–319.
[2] Liu, M.et al. *J. Nature* 2013, 501, 395
[3] Arora N, Dar M I, Hinderhofer A, Pellet N, Schreiber F, Zakeerudin S M, Grätzel M 2017 *Science* doi: 10.1126/science.aam5655
[4] Slotcavage D J, Karunadasa H I, McGehee M D 2016 *ACS Energy Lett.* **1** (6) 1199-205
[5] Hu M, Bi C, Yuan Y, Bai Y, Huang J 2016 *Adv. Sci.***3** 1500301
[6] Indari E D, Wungu T D K, Hidayat R 2017 *J. Phys.: Conf. Ser.* **877** 012054
[7] Methefessel M and Paxton A T 1989 *Physical Review B* **40** 3616-21
[8] Baikie T, Fang Y, Kadro J M, Schreyer M, Wei F, Mhaisalkar S G, Graetzel M and White T J 2013 *J. Mater. Chem. A* 5628-41
[9] Noh J H, Im S H, Heo J H, Mandal T N, Seok S I 2013 *Nano Lett.* **13** 1764-69
[10] Chang Y H, Park C H 2004 Journal of Korean Physical Society **44** 889-93
[11] H. B. Schlegel and M. J. Frisch 1991 Theoretical and Computational Models for Organic Chemistry NATO-ASI Series C vol. 339, ed J. S. Formosinho, I. G. Csizmadia, and L. G. Arnaut (Kluwer Academic, The Netherlands,) p 5-33
[12] You J, Meng L, Song T B, Guo T F, Yang Y (M), Chang W-H, Hong Z, Chen H, Zhou H, Chen Q, Liu Y, De Marco N, Yang Y 2016 *Nature Nanotechnology* **11** 75-81
[13] Rajamanickam N, Kumari S, Vendra V K, Lavery B W, Spurgeon J, Druffel T, Sunkara M K 2016 *Nanotechnology* **27** 235404
[14] Hao F, Stoumpos C C, Cao D H, Chang R P H, Kanatzidis M G 2014 *Nature Photonics* **8**(6) 489-94
[15] Wang Y, Gould T, Dobson J F, Zhang H, Yang H, Yao X and Zhao H 2014 *Phys. Chem. Chem. Phys* **16** 1424–29

IOP Conf. Series: Journal of Physics: Conf. Series **1057** (2018) 012005 doi:10.1088/1742-6596/1057/1/012005

The Effect of Growth Temperature on The Characteristics Of ZnO Nanorods And Its Optical Properties

Nandang Mufti[1,2*], Ifa K R Laila[1], Riris Idiawati[1], Abdulloh Fuad[1,2], Arif Hidayat[1], Ahmad Taufiq[1,2], Sunaryono[1,2]

[1]Department of Physics, Faculty of Mathematics and Natural Sciences, Universitas Negeri Malang (State University of Malang), Jl. Semarang No. 5, Malang 65145, Indonesia

[2]Central Laboratory for Mineral and Advanced Materials, Faculty of Mathematics and Natural Sciences, Universitas Negeri Malang (State University of Malang), Jl. Semarang No. 5, Malang 65145, Indonesia

*Corresponding author: nandang.mufti.fmipa@um.ac.id

Abstract. The aim of this research is to identify the effects of growth temperature on the characteristics of ZnO nanorods synthesized by the hydrothermal method. The Zinc oxide (ZnO) nanorods synthesis was carried out through two steps, consisting of the ZnO seed layer deposition and the ZnO nanorods growth. The ZnO seed layer was prepared by a spin coating method using zinc acetate dehydrate and ethanol on ITO substrate. The following ZnO nanorods growth was performed with the modified hydrothermal method by using hexamethylenetetramine (HMT) as a precursor and zinc nitrate with a molar ratio of 1:1 for 6 hours. A variation of ZnO nanorods growth temperatures of HMT 25 mM and 50 mM was performed in this research. The resulted ZnO nanorods were characterized by X-Ray Diffraction (XRD) and Scanning Electron Microscopy (SEM) to study their structures and morphology. Transform infrared spectroscopy (FTIR) and ultraviolet–visible spectroscopy (UV-Vis) were utilized to characterize the optical characteristics of ZnO nanorods. In general, the synthesized ZnO nanorods were found to be oriented on the (100) and (101) planes with diameters ranging from 81 nm until 365 nm and rods lengths up to 5.1 μm. The HMT 50 mM concentration showed that the higher the growth temperature, the larger the sizes of diameter and length of the ZnO nanorods. The ZnO nanorods synthesized by using the 50 mM HMT concentration at 80 oC presented the most favorable result since it had a small size, diameter, and crystallite size with a large energy gap.

Introduction

Zinc oxide (ZnO) is one of semiconducting materials that have drawn the attention of researchers due to its many possible applications, such as light-emitting diodes [1], gas sensor [2], fluid sensor [3], photocatalyst [4], solar cell [5], nanogenerator [6], and so forth. ZnO is a type-n semiconductor which is unique since it has a wide energy band gap at x (3.37 eV) and a large excitonic binding energy, namely 60 meV at room temperature [7]. The synthesis of ZnO in nanoscale offers better electronic and optical properties compared to ZnO in bulk [8]. Several nano-scaled forms of ZnO morphology have been successfully synthesized such as nanofiber [9], nanotube [10], nanowires [11], nanorods [12], and

IOP Conf. Series: Journal of Physics: Conf. Series **1057** (2018) 012005 doi:10.1088/1742-6596/1057/1/012005

nanoflower [13]. The synthesis of ZnO nanorods attracts many researchers since they can be used as resources of renewable energy for example in a solar cell or nanogenerator [6,14].

Some synthesis methods to produce ZnO nanorods that have been implemented include, among other, chemical bath deposition [15], sonochemical [16], sol-gel [17], hydrothermal [18], electrodeposition [19], chemical vapor deposition [20] and vapor-liquid-solid (VLS) [21]. The methods that use vapor phase have some drawbacks, such as the needs of a high temperature, expensive operational cost, and relatively long time [22]. One of the easy methods is a hydrothermal method. This method is simpler compared with other methods that also utilize fluid. Besides, it is easier to control the growth parameter of ZnO in order to obtain the expected form and size via this method [23]. However, this method needs particular vessel and small dimension of sample. Therefore, it's important to simplify this method. In this research, we investigated the influence of growth temperature on the morphology and optical properties of ZnO nanorods in two different precursors synthesized by the modified hydrothermal method.

Experimental Method

The method used for ZnO nanoparticle deposition was sol-gel spin coating method [24]. The material employed for the synthesis of ZnO seed layer was *zinc acetate* ($Zn(CH_3COO)_2 \cdot 2H_2O$ 0,2 M) produced by dissolving 0.878 gram of zinc acetate dehydrate into 20 ml ethanol. The prepared solution was then stirred by using a magnetic stirrer at 70 °C for 2 hours with an equal moral ratio between the (Zn(CH3COO)2.2H2O) and monoethanolamine (MEA). Such solution was then used in the formation of thin films via spin coating method at 3000 rpm for 20 seconds. The coated substrate was then heated at a temperature of 150 °C for 10 minutes to be subsequently continued with the annealing process at 550 °C for 2 hours. 25 mM and 50 mM zinc nitrate tetrahydrate ($Zn(NO_3)_2 \cdot 4H_2O$) and hexamethylenetetramine (HMTA) solutions with an equal molar ratio were put in the beaker glass and utilized for the ZnO nanorods growth. The prepared solutions were then stirred by using a magnetic stirrer for 45 minutes at room temperature. Then, the ZnO seed layers were put into the solutions with their surfaces facing upward and the baker glass was covered with plastic wrap. During the process of growth, the solutions were heated at a variation of heating temperatures of 80 °C, 90 °C, and 100 °C for 6 hours. The samples were subsequently rinsed with DI-Water, dried, and annealed at a temperature of 550 °C for 2 hours. Scanning Electron Microscopy (FEI type INSPECT-S50) was used in ZnO nanorods characterization. X- Ray Diffraction (PHILIPS type X'Pert PRO), Fourier-Transform Infrared Spectroscopy (FTIR), and UV-Visible spectrophotometer were employed to characterize the morphology, structural properties, and optical properties of the ZnO nanorods, respectively.

Results and Discussion

Figure 1 presents the SEM result on ZnO nanorods synthesized at varied temperatures in two different precursor concentrations, namely 25 mM (figure 1.a – 1.c) and 50 mm (Figure 1.d – 1.f). All the ZnO nanorods that have been synthesized were in a hexagonal form having diameter varied from 81 nm until 365 nm. In the 25 mm precursor concentration, the nanorods diameter reached its maximum value of 92 nm at 90 °C. It resulted from the increasing temperature of the solution from 80 °C to 90 °C that produced adequate heat energy for the atom nucleation so that it could improve the surface mobility to be more active [25]. The ZnO nanorods morphology was smaller at the temperature of 100 °C (Figure 2b) than at the temperature of 90 °C (Figure 1.c). It was possible because an increasing temperature would improve the aggregation and accelerate the collision of atomic nucleation during the solvothermal process that can cause dispersion and desorption on the atomic nucleation [25,26]. On the 50 mM concentration, it can be seen that (figure 1.d – 1.f) the higher the immersion temperature, the larger the nanorods diameter. Generally, the ZnO nanorods synthesized from 25 mM zinc nitrate concentration have smaller diameters than those synthesized from 50 mM zinc nitrate concentration. Such fact shows that the immersion temperature and precursor concentration strongly influence the size of ZnO nanorods as found in hydrothermal method [18,25,26].

IOP Conf. Series: Journal of Physics: Conf. Series **1057** (2018) 012005 doi:10.1088/1742-6596/1057/1/012005

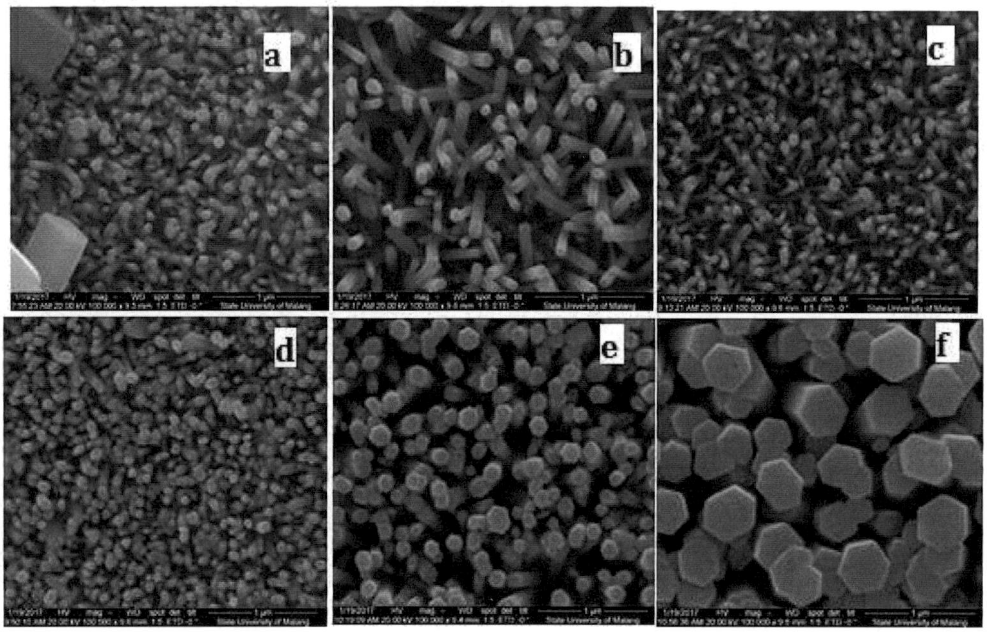

Figure 1 Morphology of ZnO nanorods synthesized for 6 hours from 25 mM zinc nitrate concentration at temperatures of (a) 80 °C, (b) 90 °C, and (c) 100 °C and 50 mm zinc nitrate concentration at temperatures of (d) 80 °C, (e) 90 °C, and (f) 100 °C.

The length of ZnO nanorods synthesized from 50 mM zinc nitrate concentration in a variation as shown in figure 2. A higher immersion temperature resulted in longer nanorods in the lengths of 2.2 μm, 4.8 μm, and 5.1 μm at the immersion temperatures of 80 °C, 90 °C, and 100 °C. It was possible since an increased temperature slowed the ZnO nucleation rate and increased the growth of $Zn(OH)_2$ that was a compound in the formation of the ZnO rods [21,26].

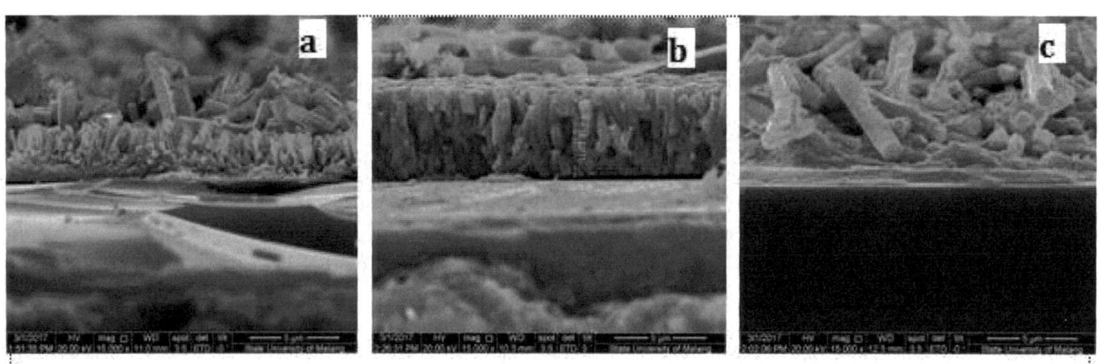

Figure 2 SEM Cross sections of ZnO NRs in 50 mM concentration at temperatures of (a) 80 °C, (b) 90 °C, and (c) 100 °C after for 6-hour synthesis.

The XRD patterns of the ZnO seed layer and nanorods at a variation of immersion temperatures in different concentrations shown in Figure 3. In general, all samples showed a single-phase ZnO with a hexagonal crystal structure (wurtzite) in accordance with the JCPDS standard No. 36-1451. The diffraction peak positions of all samples were in accordance with JCPDS although the values were slightly lower. The lattice parameters of ZnO nanorods and ZnO bulk from JPCDS were alike, namely a=3.2498 Å and c=5.2066 Å. The lattice parameters of ZnO seed layer and ZnO nanorods are tabulated in Table 1. The lattice parameter indicated that the ZnO nanorods had a crystalline form and a single phase. The peak (101) was seen to have a higher intensity compared with other peaks, except for the ZnO nanorods at 90 °C from 50 mM precursor concentration that had the highest peak (002). It indicated that the ZnO nanorods grew along the –c direction, while the peak (101) was caused by the growth of some nanorods that was not perpendicular to the substrate plane. The ZnO nanorods orientation on the plane (002) was similar to some research findings reported by other researchers [24,25]. A high immersion temperature would increase the height of the peak on (101) since the diameter and size of the nanorods were getting bigger. Such results were consistent with the SEM results.

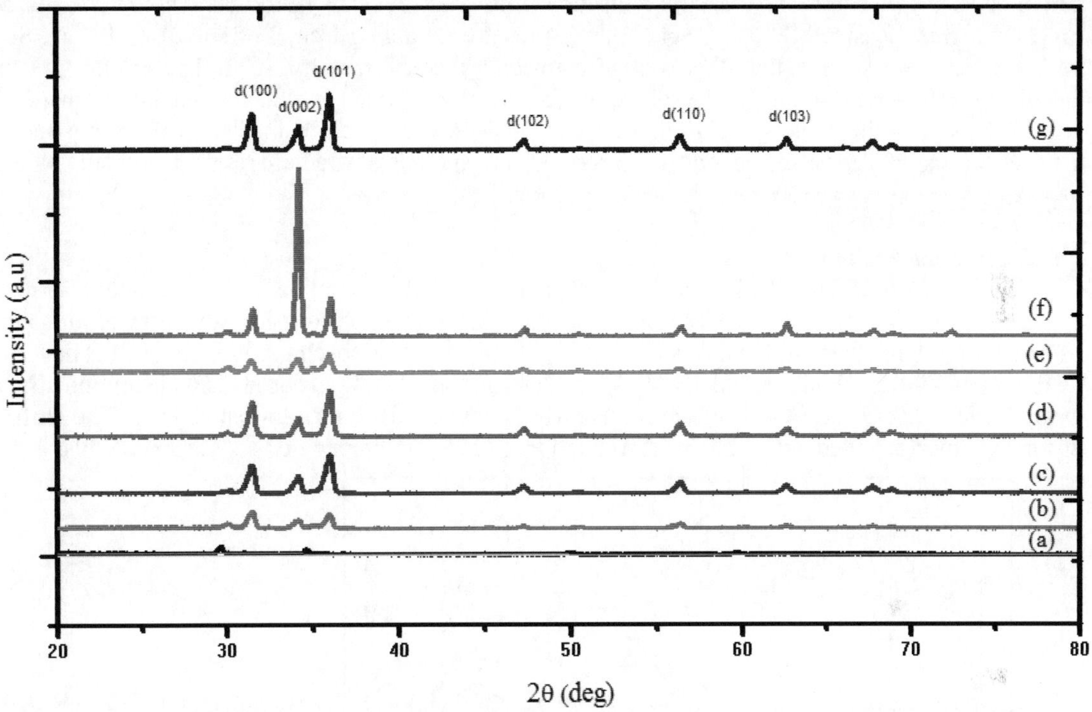

Figure 3. Diffraction patterns of (a) ZnO seed layer, ZnO nanorods at temperatures of (b) 80 °C (c) 90 °C, and (d) 100 °C from 25 mM precursor concentration, ZnO nanorods at temperatures of (e) 80 °C, (f) 90 °C, (g) 100 °C from 50 mM precursor concentration.

The average crystallite size was measured via Scherrer equation by using the FWHM value obtained from Gaussian fitting on the diffraction peak by Origin 8 software.

$$D = \frac{0,9\lambda}{B\cos\theta} \quad (1)$$

Where κ is the Scherrer constant, which depends on the crystal form and can be deemed as 0.9 [22]; λ represents the X-ray wavelength from the incident of Cu Kα radiation, which is 0.154056 nm; B is the FWHM value that is the highest full-width half value; and θ is the angle of the diffraction peak in rad. The ZnO nanorods crystallite size in Table 3 was measured based on the FWHM on the peak (101) and (002).

Table 1. Lattice parameter, crystallite size, and structure of ZnO NRs characterized by using XRD from different concentrations of zinc nitrate

	a (Å)	*c* (Å)	Crystallite size (nm)	Average diameter [nm] from SEM
ZnO NRs 25 mM 80 °C	3.2740	5.2421	17.81	81
ZnO NRs 25 mM 90 °C	3.2462	5.2018	17.34	92
ZnO NRs 25 mM 100 °C	3.2491	5.2054	19.61	81
ZnO NRs 50 mM 80 °C	3.2572	5.2139	19.76	91
ZnO NRs 50 mM 90 °C	3.2573	5.2053	21.46	148
ZnO NRs 50 mM 100 °C	3.2490	5.2040	27.19	365

According to the results shown in Table 2, it can be seen that the higher the immersion temperature and precursor concentration, the larger the crystallite size of ZnO nanorods. The ZnO nanorods crystallite sizes measured by using Scherrer equation were ranging from 17.81 until 27.19 nm. Such values were much smaller than the average diameters characterized by SEM. The crystal size that calculated by Scherer is assumed in a sphere shape. Therefore it is not applicable for nanorod shape. for However, both of them showed a similar trend. The crystallite size obtained from the nanostructure affected the energy gap value. Generally, a larger size of nanocrystal structure results in a smaller energy gap value since its surface area would become smaller [27].

FTIR Characterization

Figure 4 illustrates the FITR spectra of ZnO nanorods with a variation of immersion temperatures at various precursor concentrations. The FITR characterization was performed with the absorption areas of materials ranging from 400 until 2000 cm^{-1}. Generally, the FITR spectra references of ZnO nanorods were around 400 cm^{-1} and 500 cm^{-1} [24]. All of the samples of ZnO nanorods had absorption spectra around 400 and 485 cm^{-1} until 562 cm^{-1} that were obtained from the ZnO stretching. The shifting absorption spectra linked with a blueshift resulted from different preparations of ZnO nanorods.

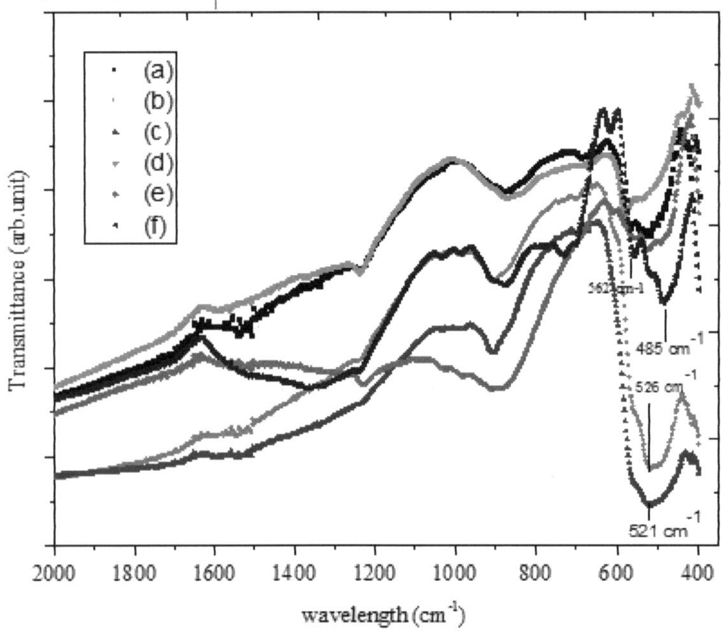

Figure 4. FTIR Spectra of ZnO NRs from 25 mM concentration at temperatures of (a) 80 °C (b) 90 °C, and (c) 100 °C and from 50 mm concentration at temperatures of (d) 80 °C, (e) 90 °C, and (f) 100 °C for 6 hours

UV-Vis Characterization

The energy gap (Eg) values were gathered by using the Tauc method from the characterization results of UV-Visible Spectroscopy. The transmittance values resulted from the UV-Vis characterization were converted into absorption coefficient (α) value by using equation 2.

$\alpha = \ln(1/T)/d$ (2)

Where T is Transmittance value of ZnO nanorods and d is film thickness. The identification of energy band-gap was performed by plotting the curve between $(\alpha h\nu)^2$ and ($h\nu$) known as Touch graph as presented in figure 5. The energy band-gap values (E_g) of ZnO nanorods are summarized in Table 3, which resulted from the intercepts of linear equation in the transitional area. The E_g values of the ZnO nanorods are shown in Table 4. The ZnO nanorods grew from 50 mM precursor concentration showed that the higher the immersion temperature, the smaller the energy band-gap, namely the E_g changed from 3.24 eV to 3.15 eV at 80 °C and 100 °C, respectively. The E_g values of ZnO nanorods were commonly showing lower redshifts approximately ranged 0.15 eV until 0.22 eV than the E_g value of bulk ZnO, of 3.37 eV. It related to the optical confinement effect in the formation of ZnO nanorods [24].

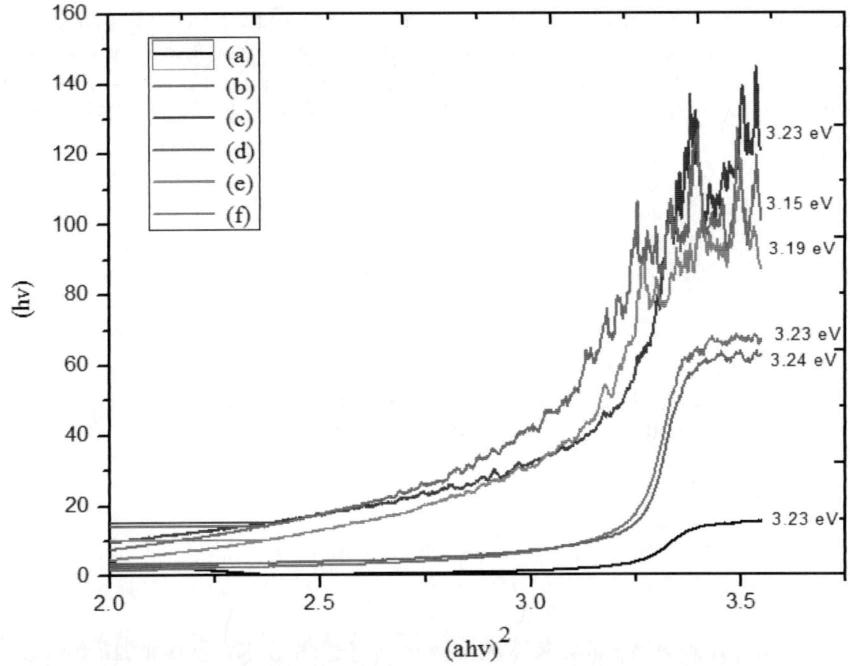

Figure 5. Graph of relations among ($h\nu$) as abscissa, $(\alpha h\nu)^n$, and energy band-gap values and a variation of ZnO NRs from 25 mM precursor concentration synthesized at temperatures of (a) 80 °C, (b) 90 °C, and (c) 100 °C and 50 mM precursor concentration synthesized at temperatures of (d) 80 °C, (e) 90 °C, and (f) 100 °C for 6 hours.

A further analysis was performed to obtain the refractive index value (n) and optical dielectric constant (ε_∞) that are summarized in Table 2. The refractive index value was measured by using equation 3 proposed by Revinda et al. [28], namely

$n = \alpha + \beta . E_g$ (3)

where the values of α = 4.048 eV^{-1} and β = -0.62 eV^{-1}, while the optical dielectric constant (ε_∞) was measured by using $\varepsilon_\infty = n^2$ [29]. Table 3 shows that the refractive index value (n) and optical dielectric constant (ε_∞) increased along with the growing immersion temperature.

Table 2 Energy band-gap of ZnO nanorods

	Band-gap (eV)	Refractive index (n)	Optical Constant (ε_∞)
Seed Layer ZnO	3.22	2.051	4.209
ZnO NRs 25 mM 80 °C	3.23	2.045	4.184
ZnO NRs 25 mM 90 °C	3.23	2.045	4.184
ZnO NRs 25 mM 100 °C	3.23	2.045	4.184
ZnO NRs 50 mM 80 °C	3.24	2.0392	4.158
ZnO NRs 50 mM 90 °C	3.19	2.0702	4.286
ZnO NRs 50 mM 100 °C	3.15	2.095	4.389

Conclusions

In this study, the ZnO NRs were synthesized by a simple hydrothermal method with a variation of zinc nitrate concentrations and at different growth temperatures. The SEM showed that the diameters of ZnO NRs ranging from 81-365 nm. The XRD diffraction values presented that the ZnO NRs had a good crystal quality and were oriented on the plane (101) and (002). The energy band-gap values of ZnO NRs synthesized from a higher zinc nitrate concentration and at increased temperature showed band-gap values that were lower than the ZnO database in general. The crystallite size, diameter, rod length, and band gap were affected by the amount of the zinc nitrate concentration and the growth temperature. The ZnO NRs synthesized from 50 mM zinc nitrate at 80 °C gave the best results regarding the structural and optical properties; ZnO NRs from 50 mm AT 80 °C showed a small diameter size, crystallite size, and showed the highest energy band-gap value. The method developed in this research offers a simple and cheap approach for producing ZnO NRs with the expected characteristics so that they could be used for optoelectronic applications.

Acknowledgment

The authors would like to thank DRPM, *Ministry of Research, Technology and Higher Education (Ristekdikti)* Republic of Indonesia for supporting a research grant for NM

References

[1] Jeong H, Salas-Montiel R, Lerondel G and Jeong M S 2017 Ultraviolet, blue, and green InGaN-based light-emitting diodes functionalized with ZnO nanorods *J. Alloys Compd.* **708** 612–8

[2] Sankar ganesh R, Navaneethan M, Mani G K, Ponnusamy S, Tsuchiya K, Muthamizhchelvan C, Kawasaki S and Hayakawa Y 2017 Influence of Al doping on the structural, morphological, optical, and gas sensing properties of ZnO nanorods *J. Alloys Compd.* **698** 555–64

[3] Water W, Chen S-E, Meen T-H and Ji L-W 2012 ZnO thin film with nanorod arrays applied to fluid sensor *Ultrasonics* **52** 747–52

[4] Kumar R, Kumar G, Al-Dossary O and Umar A 2015 ZnO nanostructured thin films: Depositions, properties and applications—A review *Mater. Express* **5** 3–23

[5] Wu S, Li J, Lo S-C, Tai Q and Yan F 2012 Enhanced performance of hybrid solar cells based on ordered electrospun ZnO nanofibers modified with CdS on the surface *Org. Electron.* **13** 1569–75

[6] Al-Ruqeishi M S, Mohiuddin T, Al-Habsi B, Al-Ruqeishi F, Al-Fahdi A and Al-Khusaibi A Piezoelectric nanogenerator based on ZnO nanorods *Arab. J. Chem.*

[7] Wahab H A, Salama A A, El-Saeid A A, Nur O, Willander M and Battisha I K 2013 Optical, structural and morphological studies of (ZnO) nano-rod thin films for biosensor applications using sol gel technique *Results Phys.* **3** 46–51

[8] Zhao L-H, Zhang J and Sun S-Q 2012 Stable aqueous ZnO nanoparticles with green photoluminescence and biocompatibility *J. Lumin.* **132** 2595–8

[9] Li F, Wang G, Jiao Y, Li J and Xie S 2014 Efficiency enhancement of ZnO-based dye-sensitized solar cell by hollow TiO2 nanofibers *J. Alloys Compd.* **611** 19–23

[10] Tsui L and Zangari G 2014 Modification of TiO2 nanotubes by Cu2O for photoelectrochemical, photocatalytic, and photovoltaic devices *Adv. Electrochem. Mater. Manuf. Pap. 13th ISE Top. Meet.-10 April 2013 Pretoria South Afr.* **128** 341–8

[11] Chappaz-Gillot C, Berson S, Salazar R, Lechêne B, Aldakov D, Delaye V, Guillerez S and Ivanova V 2014 Polymer solar cells with electrodeposited CuSCN nanowires as new efficient hole transporting layer *Sol. Energy Mater. Sol. Cells* **120, Part A** 163–7

[12] Liu Y and Gao W 2015 Growth process, crystal size and alignment of ZnO nanorods synthesized under neutral and acid conditions *J. Alloys Compd.* **629** 84–91

[13] Kim S-K, Park S, Son M-K and Kim H-J 2015 Ammonia treated ZnO nanoflowers based CdS/CdSe quantum dot sensitized solar cell *Electrochimica Acta* **151** 531–6

[14] Son D-Y, Im J-H, Kim H-S and Park N-G 2014 11% Efficient Perovskite Solar Cell Based on ZnO Nanorods: An Effective Charge Collection System *J. Phys. Chem. C* **118** 16567–73

[15] Taunk P B, Das R, Bisen D P, Tamrakar R K and Rathor N 2015 Synthesis and optical properties of chemical bath deposited ZnO thin film *Karbala Int. J. Mod. Sci.* **1** 159–65

[16] Rayathulhan R, Sodipo B K and Aziz A A 2017 Nucleation and growth of zinc oxide nanorods directly on metal wire by sonochemical method *Ultrason. Sonochem.* **35, Part A** 270–5

[17] Yoon Y-C, Park K-S and Kim S-D 2015 Effects of low preheating temperature for ZnO seed layer deposited by sol–gel spin coating on the structural properties of hydrothermal ZnO nanorods *Thin Solid Films* **597** 125–30

[18] Hassanpour A, Bogdan N, Capobianco J A and Bianucci P 2017 Hydrothermal selective growth of low aspect ratio isolated ZnO nanorods *Mater. Des.* **119** 464–9

[19] Hou Q, Zhu L, Chen H, Liu H and Li W 2013 Highly regular and ultra-thin porous ZnO nanosheets: An indirect electrodeposition method using acetate-containing precursor and their application in quantum dots-sensitized solar cells *Electrochimica Acta* **94** 72–9

[20] Ye Z, Wang T, Wu S, Ji X and Zhang Q 2017 Na-doped ZnO nanorods fabricated by chemical vapor deposition and their optoelectrical properties *J. Alloys Compd.* **690** 189–94

[21] Alvi N H, Hassan W ul, Farooq B, Nur O and Willander M 2013 Influence of different growth environments on the luminescence properties of ZnO nanorods grown by the vapor–liquid–solid (VLS) method *Mater. Lett.* **106** 158–63

[22] Rezabeigy S, Behboudnia M and Nobari N 2015 Growth of ZnO Nanorods on Glass Substrate by Chemical Bath Deposition *5th Int. Bienn. Conf. Ultrafine Grained Nanostructured Mater. UFGNSM15* **11** 364–9

[23] Ullah S, Ahmed F, Badshah A, Ali Altaf A, Raza R, Lal B and Hussain R 2013 Solvothermal Preparation of ZnO Nanorods as Anode Material for Improved Cycle Life Zn/AgO Batteries ed B Xu *PLoS ONE* **8** e75999

[24] Foo K L, Hashim U, Muhammad K and Voon C H 2014 Sol–gel synthesized zinc oxide nanorods and their structural and optical investigation for optoelectronic application *Nanoscale Res. Lett.* **9** 429

[25] Meng Y, Lin Y and Yang J 2013 Synthesis of rod-cluster ZnO nanostructures and their application to dye-sensitized solar cells *Appl. Surf. Sci.* **268** 561–5

[26] Shaziman S, Ismail@rosdi A S, Mamat M H and Zoolfakar A S 2015 Influence of Growth Time and Temperature on the Morphology of ZnO Nanorods via Hydrothermal *IOP Conf. Ser. Mater. Sci. Eng.* **99** 012016

[27] Lin H, Huang C, Li W, Ni C, Shah S and Tseng Y 2006 Size dependency of nanocrystalline TiO2 on its optical property and photocatalytic reactivity exemplified by 2-chlorophenol *Appl. Catal. B Environ.* **68** 1–11

[28] Ravindra N M, Auluck S and Srivastava V K 1979 On the Penn Gap in Semiconductors *Phys. Status Solidi B* **93** K155–60

[29] Samara G A 1983 Temperature and pressure dependences of the dielectric constants of semiconductors *Phys. Rev. B* **27** 3494–505

IOP Conf. Series: Journal of Physics: Conf. Series **1057** (2018) 012006 doi:10.1088/1742-6596/1057/1/012006

Non-iterative model based image reconstruction of diffuse optical tomography based on DE and RTE in quality control of agricultural product studies

Vebi Nadhira[1*], E. Juliastuti[1], Agah D. Garnadi[2], Deddy Kurniadi[1], and Yoko Hoshi[3]

[1]Instrumentation and Control Research Group, Faculty of Industrial Technology, Institute Technology of Bandung Ganesha 10, +62 22 250 4424, 40132 Bandung Indonesia

[2]Departement of Mathematics, Faculty of Mathematics and Natural Sciences, Bogor Agricultural University, Jl Meranti, Kampus Darmaga, 16680, Bogor, Indonesia

[3]Department of Biomedical Optics, Medical Photonics Research Center, Hamamatsu University School of Medicine, 1-20-1 Handayama, Higashi-ku, Hamamatsu city, Shizuoka, Japan 431-3192

vebi@tf.itb.ac.id

Abstract. In a previous study, we developed the non-iterative image reconstruction based on diffusion equation. Within this research, we applied the same non-iterative algorithm scheme using radiative transfer equation. Basically, the non-iterative image reconstruction is a development of model based image reconstruction that implemented truncated singular value decomposition and L-curve analysis to solve the ill-posed problem. This algorithm reduces the computation time to reconstruct the cross-sectional area. As part of continuous development on agricultural application, optical tomography based quality control is offered. An experiment was conducted on potato to evaluate these two non-iterative algorithms. The object was illuminated by the near infrared source from 8 positions on the object boundary. In this experiment, we vary the position and the amount of epoxy on the object then we analyze the residual value between measurement and reconstructed boundary data. The reconstructions were performed in continuous-wave domain. Furthermore, we compared the residual value from diffuse optical tomography using diffusion equation and radiative transfer equation. The result of this study indicates that these algorithms are promising to detect the presence of epoxy on potato which is useful for quality control of agricultural products.

1. Introduction

Several researchers have been developing the reconstruction methods to reduce the computation time in optical tomography (OT) with the regularization parameter selection method [7], a fast-forward solver of radiative transfer equation (RTE) [5], non-iterative algorithm [8], hardware development [14], and signal processing approach [1, 13, 15]. This paper focused on the study of non-iterative image reconstruction based on diffusion equation (DE) and RTE. To avoid the iterative step, L-curve based algorithm for automatic selection of optimal k value is performed to obtain truncated singular value decomposition (TVSD) matrix which will derive a new well-posed problem. In forward solving, the continuous wave signal processing approach was applied in order to get a faster boundary data calculation. Furthermore, a fast RTE solver that has been developed by H. Gao and H. Zhao was applied to solve the forward problem in OT based on RTE.

Usually, the development of non-destructive evaluation (NDE) based on optical tomography method has been focused on medical application [4, 6, 8, 9]. Nevertheless, NDE OT development on other application is also interesting and promising as described by some researcher on this field [3, 14]. This paper as well as our previous papers mainly study NDE based on OT method for agricultural product quality control [17, 18, 19]. Therefore, the experiment of internal defects monitoring on agricultural products using OT is described on this paper to validate our proposed non-iterative image reconstruction algorithm.

2. NIMoBIR OT (Non-Iterative Model Based Image Reconstruction of Diffuse Optical Tomography)

2.1. Principle of Optical Tomography (OT)

OT reconstruction system is divided into two essential parts. The first one is forward problem solving and the other one is inverse problem solving [13, 20], as shown in figure 1. The forward problem solving is used to calculate the boundary data of prediction model and/or simulation object when the optical properties of the medium and the amount of input light are given. In this paper, we used DE and RTE to describe the light propagation in biological material, which then are used as forward problem solving [1, 5, 16]. While inverse problem solving is an algorithm to predict an optical coefficient distribution inside a medium by comparing the boundary data of the "object" and those of prediction model.

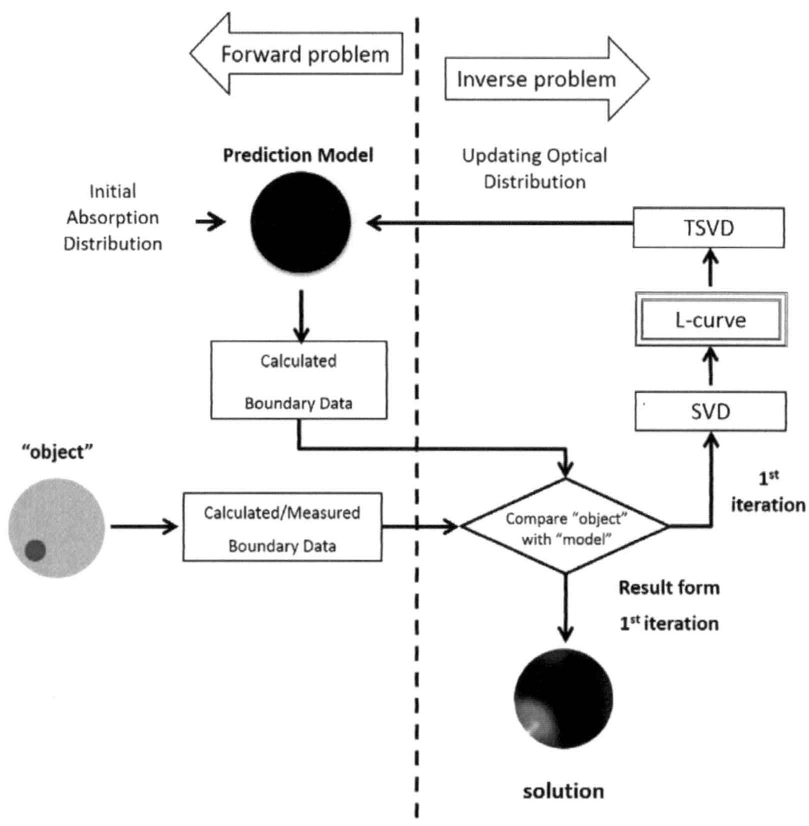

Figure 1. NIMoBIR algorithm

A fundamental model for light transport in tissues is the radiative transfer equation. The continuous wave domain RTE is of the form

$$\hat{s}.\nabla.L(r,\hat{s})+(\mu_a+\mu_s)L(r,\hat{s})=\mu_s\int_{S^{n-1}} f(\hat{s},\hat{s}')L(r,\hat{s})d\hat{s}'+q(r,\hat{s}) \tag{1}$$

where μ_a and μ_s are absorption and scattering coefficient of the medium, respectively. Furthermore, $L(r,\hat{s})$ is the radiance and $f(\hat{s}.\hat{s}')$ is the probability density function. The probability density function describes the probability of scattering from direction $\hat{s}$ into direction $\hat{s}'$. $q(r,\hat{s})$ is the source per unit volume at position r along $\hat{s}$. Even though RTE is a widely accepted model, DE is more commonly used because of its computational simplicity. In DE framework, the RTE is approximated as follows:

$$-\kappa\nabla^2.\Phi(r)+\mu_a\Phi(r)=q_0(r) \tag{2}$$

where Φ is the diffuse photon fluency rate, q_0 is the integrated light source term, and κ is the optical diffusion coefficient [1, 2, 15]. The optical diffusion coefficient can be represented as follows:

$$\kappa=\frac{1}{3[\mu'_s+\mu_a]} \tag{3}$$

where μ'_s [m^{-1}] is the reduced scattering coefficient [13,16].

In order to simplify the boundary data calculation process using forward problem, discretization approach needs to be applied. In this paper FEM approach is used to divide the object or model into non-overlapping triangle elements as shown in figure 2.

Furthermore, FEM approach is applied to simulation object and prediction model to simplify the computational process. This FEM approach is dividing the object or model into non-overlapping triangle elements in order to discretize the diffusion equation as shown in figure 2.

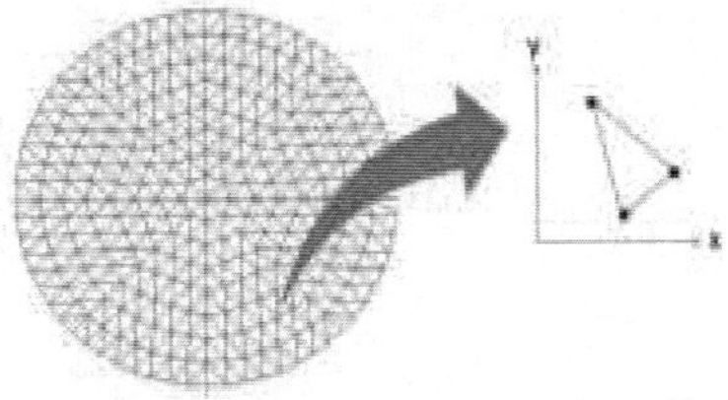

Figure 2. Discretization object with FEM [20]

FEM is selected because it can provide fast, accurate and flexible modeling for complex geometries [17, 19]. By applying FEM to DE and RTE, the photon density Φ and the radiance L is approximated by a piecewise linear and continuous function. Definition of Φ^h and L^h are as follows:

$$\Phi^h(r)=\sum_{j=1}^{N}\Phi_j(r)u_j(r) \tag{4}$$

$$L^h(r)=\sum_{j=1}^{N}L_j(r)u_j(r) \tag{5}$$

where u_j are linear basis functions for which the values at element centers n_j are defined as $u_i(n_j)=\delta ij$ ($i,j=1,\ldots,N$). Either Φ^h or L^h will be used as input for the inverse problem.

In image reconstruction algorithm of OT, the optical coefficient distribution of the "object" or/and predictive model are estimated based on boundary data and light propagation within the medium.

IOP Conf. Series: Journal of Physics: Conf. Series **1057** (2018) 012006 doi:10.1088/1742-6596/1057/1/012006

The boundary data from both DE and RTE can be described as a general vector b and the optical coefficient distribution can be described by a general vector μ. Then, a physical model that describes the relation between b and μ can be written as follows:

$$J = \frac{\partial b}{\partial \mu} \tag{6}$$

where J is the sensitivity matrix that defines sensitivity of F to perturbations in μ, also known as Jacobian matrix. Furthermore, the forward problem is defined as finding the values of data b from given μ, for which:

$$b = J\mu$$

An image representing optical coefficient distribution μ from given data b needs to be estimated in inverse problem, which can be written as:

$$\mu = J^{-1}b \tag{7}$$

Unfortunately, the Jacobian inversion process leads to ill-posed problem. Generally, it can be solved by applying iterative reconstruction. However, in this paper TSVD is used to solve this problem in a non-iterative way. The first step in TSVD algorithm is converting matrix J^{-1} to singular value decomposition matrix as follows:

$$J^{-1} = VSD^{T} \tag{8}$$

where S = diag(s_1, … , s_n) has non-negative diagonal elements appearing in non-increasing order [12]. In this case, $V = (v_1, \dots, v_n)$ and $D = (d_1, \dots, d_n)$ are matrices with orthonormal columns. Furthermore, to derive a new problem with well-posed *rank deficient* coefficient matrix, we need to truncate the S matrix.

The rank deficient matrix, S_k, is obtained by truncating the S at k so that S_k is the best rank-k approximation to S. S_k is given by

$$S_k = \sum_{i=1}^{k} s_i \,,\; k \le n \tag{9}$$

L-curve analysis is the most convenient graphical tool for determining the optimum value of k precisely [11]. L-curve plot the norm $\|\mu\|_2$ of the SVD solution to the prediction error norm $\|S_k\mu-b\|_2$, where k minimizes these two quantities at the same time.

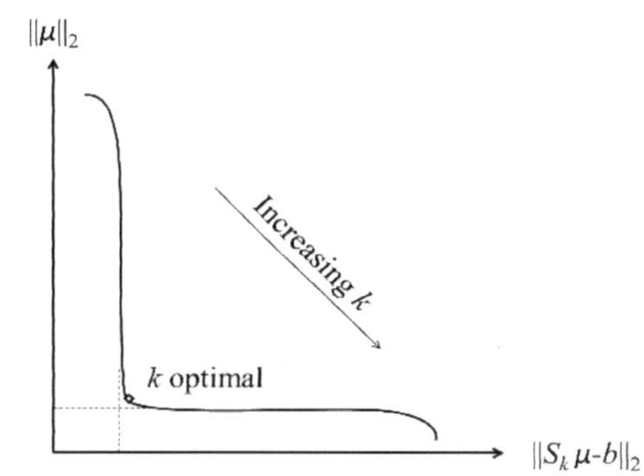

Figure 3. L-curve analysis for TSVD

3. Experiment of potato monitoring

Evaluation of the proposed non-iterative model based image reconstruction method was performed on a ring array optical tomography system used for quality control of potato. Measurement was done using cylindrical carrot with diameter and height of $30*10^{-3}$ m and $40*10^{-3}$ m, respectively. The experiment was done on four different circular objects with several defect conditions as shown in figure 5.

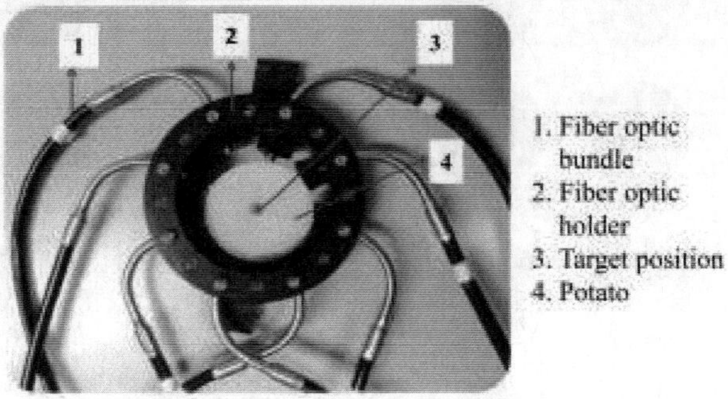

Figure 4. Experimental configuration

The experimental configuration is shown in figure 4. In this experiment, near-infrared laser emits continuous light having peak powers around 0.5 W at 759 nm. A beam of light is projected to the object through a single fiber at the center of fiber optic bundle. Then the scattered and absorbed light intensity is received by a photodetector via annular fibers within fiber optic bundle.

There are 8 test points on the object boundary that will be occupied by source and detector as shown in figure 4. The boundary data matrix is denoted as:

$$b = \begin{bmatrix} b_{1,1} & b_{2,1} & \cdots & b_{8,1} \\ b_{1,2} & b_{2,2} & \cdots & b_{8,2} \\ \vdots & \vdots & \ddots & \vdots \\ b_{1,8} & b_{2,8} & \cdots & b_{8,8} \end{bmatrix}_{NSxND} \tag{10}$$

where NS indicate number of sources and ND indicate number of detectors.

4. Results and analysis

The experiment objects with various target positions and the reconstructed images are shown in **Figure 5**. General conclusion based on visual analysis is that the NIMoBIR algorithm using DE succeeded to represent the position and the size of the objects. However, there are several white spots on reconstructed images that are difficult to distinguish from the presence of the target. Furthermore, this method is unable to show the separation between two targets on object. Meanwhile, as shown in **Figure 5**, the NIMoBIR using RTE has a better ability to separate two targets on object than the NIMoBIR using DE. However, this method still has weakness in reconstructing the target position. In order to carry out an objective analysis, the residual value is used to show the difference between measurement and reconstructed boundary data, as shown in table 1.

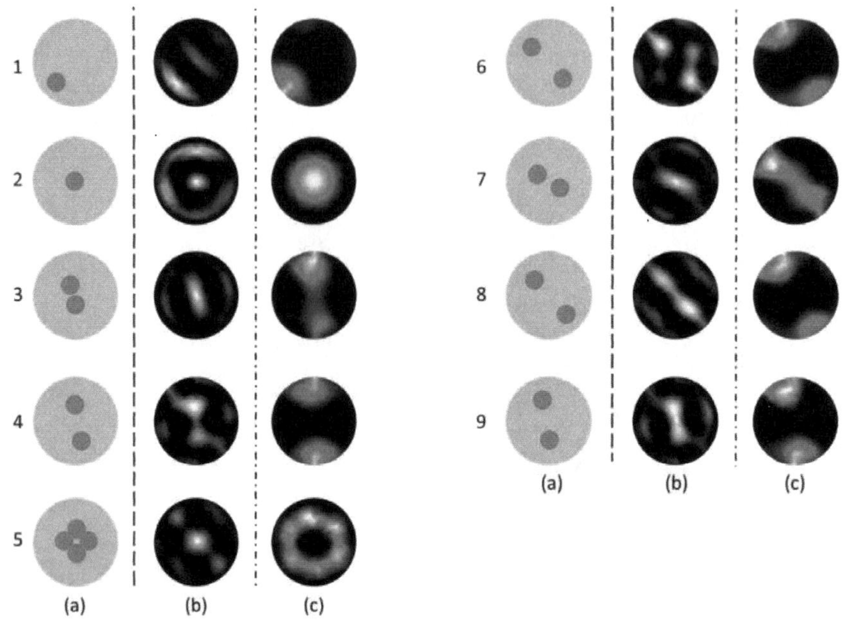

Figure 5. The results of CW-NIMoBIR reconstruction (a) the depiction of object condition (b) The reconstructed image using RTE (c) The reconstructed image using DE.

Table 1. Numerical analysis calculation results on NIMoBIR CW using DE and RTE

Object	**Residual Value**	
	DE	**RTE**
1	$8,82*10^{-7}$	$6,5*10^{-12}$
2	$1,2*10^{-6}$	$8,2*10^{-9}$
3	$1,91*10^{-6}$	$7,49*10^{-12}$
4	$5,33*10^{-6}$	$2,87*10^{-12}$
5	$4,15*10^{-6}$	$3,36*10^{-13}$
6	$2,3*10^{-6}$	$1,08*10^{-9}$
7	$2,4*10^{-6}$	$6,34*10^{-12}$
8	$1,93*10^{-6}$	$8,36*10^{-10}$
9	$2,58*10^{-6}$	$3,73*10^{-12}$

As shown in table 1, the averages of residual value on NIMoBIR CW DE and RTE are $2.52 * 10^{-6}$ and $1.13 * 10^{-9}$ respectively. These averages are relatively low, so it can be concluded that both algorithms are able to reconstruct the boundary data of the real object. The lower residual value obtained by applying NIMoBIR using RTE showed that this algorithm has a better performance to represent the measurement boundary data than the use of NIMoBIR using DE. Based on visual and residual analysis, it can be concluded that the algorithm NIMoBIR with continuous-wave domain using DE and RTE systems have the potential to reconstruct the absorption coefficient distribution of agricultural products, such as potato.

5. Conclusion

This article presents non-iterative model based image reconstruction for continuous-wave domain diffusion and radiative transport equations. Experiments using potato with the epoxy targets were conducted to demonstrate the proposed algorithm. The result of this experiment has shown that the

NIMoBIR using DE and RTE have high possibility to detect defects in agricultural products. Accuracy improvement of the reconstructed image will be conducted in future work.

References

[1] A. Marjono, A. Yano, S. Okawa, F. Gao, and Y. Yamada, "Total light approach of time domain fluorescence diffuse optical tomography", *Optics Express*, vol. 16, no. 17, pp. 13104-13121, 2008.

[2] A. Zacharopoulos, M. Schweiger, V. Kolehmainen, and S. Arridge, "3D shape based reconstruction of experimental data in Diffuse Optical Tomography", *Optics Express*, vol. 17, Issue 21, pp. 18940-18956, 2009.

[3] E. Kate K, Henri S. Tapp, Richard B, Robert O. M, and Anthony J.P, "Feasibility study of NIR diffuse optical tomography on agricultural produce", *Science Direct Postharvest Biology and Technology*, vol. 48, issue 2, pp. 223–230, 2008.

[4] H. Egger, M. Freiberger, and M. Schlottbom, 2010, "On forward and inverse models in fluorescence diffuse optical tomography", *Inverse problems and Imaging*, vol.4, no. 3, pp. 411-427,.

[5] H. Gao and H. Zhao, "A fast forward solver of radiative transfer equation", *Transport Theory and Statistical Physics*, vol. 38, pp. 149-192, 2009.

[6] Harsimrat Singh et al., "Mapping cortical haemodynamics during neonatal seizures using diffuse optical tomography: A case study", *NeuroImage: Clinical*, vol. 5, pp. 256–265, 2014.

[7] J. Chamorro et al., 2011, "Feasibility of U-curve method to select the regularization parameter for fluorescence diffuse optical tomography in phantom and small animal studies", *Optics Express*, vol. 19, Issue 12, pp. 11490-11506, 2011.

[8] Lee O., Kim J. M., Bresler Y., and Ye J.C., "Compressive Diffuse Optical Tomography Noniterative exact reconstruction using joint sparsity", *IEEE Trans. Med. Imaging*, vol. 30, issue 5, pp 1129-1142, 2011.

[9] M. L. Flexman, M. A. Khalil, R. Al Abdi, H. K. Kim, C. J. Fong, E. Desperito, D.L. Hershman, R. L. Barbour, and A. H. Hielscher, "Digital optical tomography system for dynamic breast imaging", *J. Biomed Opt,* vol. 16 issue 7: 076014, 2011.

[10] M. Panayiotou, A. Gibson, and Prof. J. Hebden, "*3Dimensional Optical Tomography of an Unfertilised Chicken Egg"*, BSc Thesis, Medical Physics, University College London, 35 p, 2005

[11] P.C. Hansen, "Discrete Inverse Problems Insight and Algorithms", *SIAM*, 2010.

[12] S. Gupta, Phaneendra K. Y, D. Roy, D. Piao, and Ram M. V., "Singular value decomposition based computationally efficient algorithm for rapid dynamic near-infrared diffuse optical tomography", *Medical Physics*, vol. 36, no. 12, pp. 5559-5567, 2009.

[13] S. Okawa et al.., "Phantom and mouse experiments of time-domain fluorescence tomography using total light approach", *Biomed.Opt. Express*, vol. 4, no.4 , pp. 635-651, 2013.

[14] S. Z.M. Muji et al., "Optical tomography hardware development for solid gas measurement using mixed projection", Flow Measurement and Instrumentation, vol. 33, pp. 110–121, 2013.

[15] Shay Keren, O.Gheysens, C. S. Levin, and S. S. Gambhir, "A comparison between a time domain and continuous wave small animal optical imaging system", *IEEE Medical Imaging*, vol. 27, no. 1, pp. 58-63, 2008.

[16] T. Tarvainen et al., "An approximation error approach for compensating for modelling errors between the radiative transfer equation and the diffusion approximation in diffuse optical tomography", *Inverse Problems,* vol. 26, 015005, 2010.

IOP Conf. Series: Journal of Physics: Conf. Series **1057** (2018) 012006 doi:10.1088/1742-6596/1057/1/012006

[17] V. Nadhira, D. Kurniadi, and E. Juliastuti, "Feasibility Study on Image Reconstruction of Continuous Wave Domain Diffuse Optical Tomography for Quality Control on Seed Potatoes", IEEE, *ICICI-BME Conference Proceeding*, pp. 421 – 424, 2013.

[18] V. Nadhira, D. Kurniadi, E. Juliastuti, and Adeline Sutiswan, "Study of continuous-wave domain fluorescence diffuse optical tomography for quality control on agricultural produce", *AIP Conference Proceedings*, vol. 1589, pp. 276-280, 2014.

[19] V. Nadhira, D. Kurniadi, E. Juliastuti, and R. Richo Eka, "Image Reconstruction of Time Domain Diffuse Optical Tomography for Quality Control on Seed Potatoes", IEEE, *ICA Conference Proceedings*, pp. 266-269, 2013.

[20] Xiaofeng Zhang, "Instrumentation in Diffuse Optical Imaging", *Photonics,* vol. 1, no. 1, pp. 9-32, 2014.

IOP Conf. Series: Journal of Physics: Conf. Series **1057** (2018) 012007 doi:10.1088/1742-6596/1057/1/012007

Preparations of Organo-Lead Halide Perovskite Layers in Humid Air Atmosphere and their Characteristics

Adhita Asma Nurunnizar, Muslihun, and Rahmat Hidayat

Physics of Magnetism and Photonics Research Division, Faculty of Mathematics and Natural Sciences, Bandung Institute of Technology, Indonesia

E-mail: rahmat@fi.itb.ac.id

Abstract: In recent years, solar cells based on organometal halide perovskite (hybrid perovskite) materials have emerged as promising low-cost solar cells. These types of solar cells exhibit remarkable high power conversion efficiencies while offering a significant advantage on the simplicity of their fabrications. The formation of organo-metal halide perovskite layer, however, is very sensitive to some process parameters such as concentration, temperature, and humidity. Here, we report the preparation of this perovskite crystal layer of $CH_3NH_3PbI_3$ ($MAPbI_3$) under relatively humid air atmosphere, which is a common condition in tropical countries, and understanding its effect on the resulted perovskite layer characteristics. We found that the humidity leads to the formation of a black color layer on the top of dark brown color perovskite layer. This black color layer caused the appearance of optical absorption tail extending to the near infrared region and showed poor x-ray diffraction (XRD) pattern. It also leads to poorer solar cell characteristics. The formation of this layer can be prevented by applying fast deposition-crystallization process with proper solvent and dropping time.

1. INTRODUCTION

Studies on organo-metal halide perovskites (OMHP) for solar cells application has significantly increased in the last five years after the reports by two different research groups, namely Snaith and Gratzel groups, who reported high power conversion efficiency (PCE) of 10% at their first attempts [1-4]. Although perovskite materials have long been known, such as $CaTiO_3$ which is a kind of perovskite oxide, they are mostly studied for their optical and magnetic properties. Although both the OMHP and the oxide perovskites have the same general crystal structure, they have different electronic properties in which the halide perovskites behave as a semiconductor material and exhibits photovoltaic characteristics. In addition to the OMHP advantages in producing solar cells with high conversion efficiency, these perovskites are also interesting because of the easiness in their preparation. They can be prepared via wet-chemical or solution processes at room temperature, which might reduce the cost of solar cell manufacturing.

The formation of these OMHP crystals is simply based on stoichiometry reaction. However, the formation process seems to be sensitively affected by some process and environmental parameters such as concentration, temperature, and humidity. It is therefore interesting to explore the preparation of these perovskite crystal layers under relatively humid air atmosphere, which is a common condition in tropical countries, and understanding the effects of this humid air on the characteristics of the resulted OMHP crystal layers and the performance of their solar cells. There are many reports on the preparation methods of these perovskites so far, but simple wet-chemical or solution processes called as One Step Precursor Deposition (OSPD) and Two Step Precursor Deposition (TSPD) are the most popular methods [1, 5-7]. It has also been reported how the perovskite layer color may vary depending on the detail variations of process parameters, temperature, and atmospheric gases [5]. Further improvement in OSPD technique is by dropping a small amount of an anti-solvent during the spin-coating process for reducing the solubility of OMHP precursor rapidly and thereby promoting fast nucleation and rapid crystallization. [8] This process is often called as the fast deposition-crystallization (FDC) process.

2. EXPERIMENTAL

In this work, the compact and mesoporous TiO_2 layers were prepared by spin coating technique on a pre-cleaned FTO substrate at 5000 rpm for 30 s as reported in our previous studies, but the thicknesses of the layers were much thinner than the previous samples [9,10]. After the spin coating process, those TiO_2 layers were annealed sequentially at 100°C for 15 min, 300°C for 15 min and 500°C for 30 min. The perovskite layers were then prepared either by the simple OSPD method or the OSPD+FDC method. The perovskite precursor solution was made from 1,383 mg of PbI_2 and 477 mg MAI in 1,800 mg DMF and 234 mg DMSO in a glovebox with air humidity of 30%-40%. This perovskite precursor solution was spin coated onto the TiO_2 layer (with the structure of m-TiO_2/c-TiO_2/FTO) by spin coating at 5000 rpm for 60 s. Diethyl-ether was used as the anti-solvent. After the spin coating, the film was annealed at 90°C for 15 min. For constructing a complete solar cell, the hole transport layer and Au electrode were sequentially deposited on the top of the perovskite layer. The resulted perovskite layers in this work were characterized by absorption (UV-Vis) spectroscopy, Scanning Electron Microscopy (SEM) and X-Ray Diffraction (XRD) measurements.

3. RESULTS AND DISCUSSIONS

Figure 1. SEM images of the $CH_3NH_3PbI_3$ perovskite layer surface prepared by (a) OSPD (b) OSPD+FDC (c) OSPD+FDC (enlarged magnification).

Figure 1 shows the SEM images of the resulted $CH_3NH_3PbI_3$ perovskite layers, showing difference surface morphologies depending on the detail preparation processes. In this humid air condition, the simple OSPD method produced a perovskite layer looks like a resemble of random needle-like shapes or an overlapping web, which is similar as reported previously. [9] It can also be seen easily the presence of the areas on the layer that are not covered by the perovskite crystals. For the samples prepared by the OSPD+FDC method, its surface morphology looks like flake-like shapes or island-like shapes, with about 300 nm - 900 nm in size, that are separated by a very narrow gap of just 10-20 nm. The formation of needle-like shapes seems more favorable in the humid air because such morphology was not found in samples prepared in dry air atmosphere as reported in literature elsewhere. [6] The resulted layer visually looks like having a thin black or greyish black color layer on the surface. This is indicated by absorption spectrum tail broadens up to the near infra-red region (up to 1100 nm) as shown in Figure 2. On the other hand, the OSPD+FDC with the anti-solvent used here seems to be less affected by the air humidity. Under perfect preparation condition and process, even in humid air condition, a dark brown color perovskite layer covering the entire surface of TiO_2

layer, as shown in Figure 1 (b) and (c), can be eventually obtained. This is also indicated by the cut off of absorption spectrum at about 780 nm as shown in Figure 2 (b). This FDC process or anti-solvent treatment can effectively reduce the formation of the greyish black color particles.

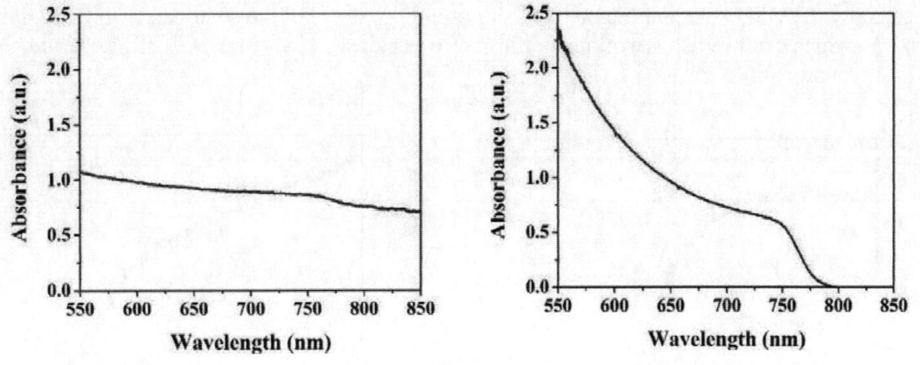

(a) (b)

Figure 2. The absorption spectra of the $CH_3NH_3PbI_3$ layer prepared by the (a) OSPD and (b) OSPD+FDC methods.

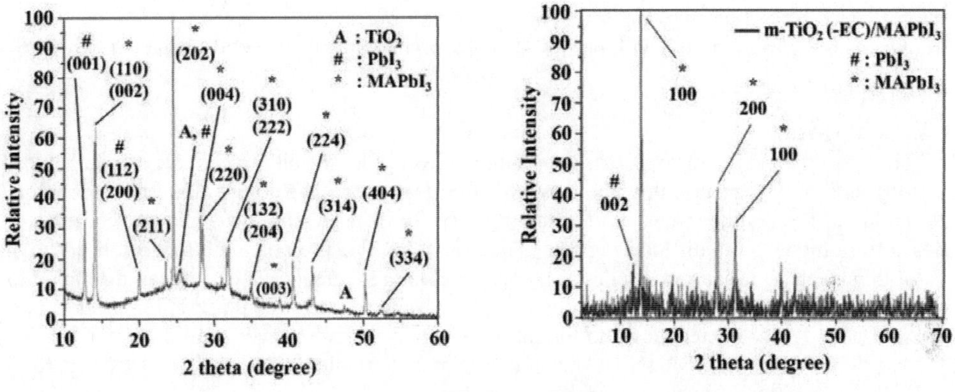

(a) (b)

Figure 3. The XRD patterns of $CH_3NH_3PbI_3$ perovskite layers prepared by the (a) OSPD and (b) OSPD+ FDC methods.

The XRD patterns of these $CH_3NH_3PbI_3$ perovskite layers are shown in Figure 3. The XRD pattern for the simple OSPD sample seems to exhibit more XRD peaks originating from large (*hkl*) planes. On the other hand, the XRD pattern for the OSPD+FDC sample shows less number of XRD peaks. The XRD peaks are in good agreement with the literature, with the most intense peak is at 13.95° originating from the (100) diffraction plane. The presence of small peaks at 28.30° and 31.79° can be assigned as the diffraction from (200) and (201) planes. In addition to those peaks, for the simple OSPD sample, there also appears a small peak that does not belong to the perovskite crystal. This peak was observed at $2\theta = 12.38°$, which can be assigned as the diffraction from the (002) plane of PbI_2. The presence of this PbI_2 peak in XRD patterns shows that there is a portion of the unconverted or remaining PbI_2 inside this perovskite layer. In contrast, this PbI_2 peak is almost negligible for the OSPD+FDC sample. The presence of greyish black or black color layer, however, reduces the height of XRD peak intensity suggesting that this layer is an amorphous phase or non-crystalline structure. It should be noted that the XRD pattern in the OSPD sample has much larger integration time and accumulation numbers.

Figure 4(a) shows the J-V curve of the perovskite solar cells (PSC) with the perovskite layer prepared by the OSPD method. Figure 4(b) shows the J-V curve of the solar cell with OSPD+FCD methods, which exhibits better working performance as indicated by a higher J_{sc} of 18.4% and better PCE of around 7.0%.

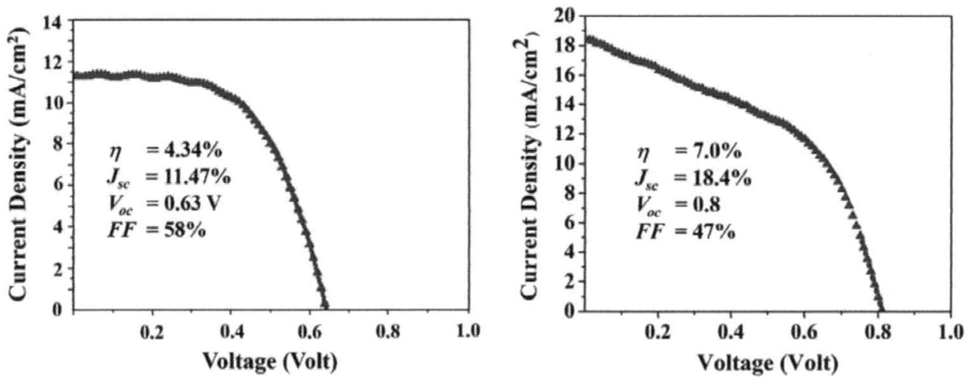

(a) (b)

Figure 4. J-V curves of PSC with perovskite layer prepared by the (a) OSPD and (b) OSPD+FDC methods

4. SUMMARY

The present experimental results show that the effect of humid air atmosphere can be reduced by employing FDC treatment following the simple OSPD process. The humid air seems to promote the formation of needle-like shape, as indicated by their SEM image and the appearance of more XRD peaks from higher (*hkl*) diffraction plane. The humid air also promotes the formation of the black color thin layer, which shows smaller XRD peak and long absorbance spectrum tail extending to the near infrared region. Because solar cells using perovskite layer with black color layer formation showed poor J-V characteristics, the formation of this black color layer should be avoided. In this work, by employing the OSPD+FDC method, a solar cell sample producing J_{sc} of 18.4% and PCE of 7.0% has been obtained even in humid air condition.

Acknowledgment

The authors acknowledge the support from Program Penelitian Unggulan Perguruan Tinggi (PUPT) ITB-Ristekdikti 2017.

References

[1] Burschka J, Pellet N, Moon S J, Baker R H, Gao P, Nazeeruddin M K, Grätzel M 2013 *Nature* **499** 316-20

[2] Liu M, Johnston M B, Snaith H J 2013 *Nature* **501** 395

[3] Lee M M, Teuscher J, Miyasaka T, Murakami T N and Snaith H J 2012 *Science* **338** 643-7

[4] Etgar L, Gao P, Xue Z, Peng Q, Chandiran A K, Liu B, Nazeerudin M K and Grätzel M 2012 *J. Am. Chem. Soc.* **134** 17396–99

[5] Dualeh A, Tétreault N, Moehl T, Gao P, Nazeeruddin M K, Grätzel M 2014 *Adv. Funct. Mater.* **24** 3250–8,

IOP Conf. Series: Journal of Physics: Conf. Series **1057** (2018) 012007 doi:10.1088/1742-6596/1057/1/012007

[6] Zheng L, Zhang D, Ma Y, Lu Z, Chen Z, Wang S, Xiao L and Gong Q, 2015 *Dalton Trans* **44** 10582-93
[7] Grätzel M 2014 *Nat. Mater* **13** 838–42
[8] Xiao M, Huang F, Huang W, Dkhissi Y, Zhu Y, Etheridge J, Weale A, Bach U, Cheng Y B and Spiccia L 2014 *Angewandte Chemie* **53** 1-7
[9] Alturisa M I, Wira J, Mardiyati, Herman and Hidayat R, 2017 *J. Phys.: Conf. Ser.* **877** 012046
[10] Dewinggih T, Shobih, Muliani L, Herman, Hidayat R, 2017 *J. Phys.: Conf. Ser.* **877** 012043

IOP Conf. Series: Journal of Physics: Conf. Series **1057** (2018) 012008 doi:10.1088/1742-6596/1057/1/012008

Size and shape effects on the field enhancement induced at a silver triangular tip

N Perdana, A A Iskandar, M O Tjia

Department of Physics, Institut Teknologi Bandung, Indonesia

iskandar@fi.itb.ac.id

Abstract. A computational study is conducted on the near field enhancement induced by localized surface plasmon resonance (LSPR) around a silver nanotape of various sizes and shapes. The results show the consistent dominance of the dipole field generated at the rectangular Ag nanotape surfaces of various lengths by a normal incident electromagnetic plane wave of polarization along the long edge. The dipole field is further shown to exhibit monotonous spectral redshifts and increased magnitudes with increased length (ℓ) of the rectangular nanotape. In contrary, the spectral peak and the field magnitude do not show the simple consistent variations with varied apex angle (θ) of the triangular nanotape. The calculated result for the triangular tape does exhibit the largely concentrated near field magnitude at the tip. It is found that a remarkable enhancement factor of 10^7 is attained at $\theta = 45°$ and $\ell = 50$ nm.

1. Introduction

The surface plasmon generated by an incident light at a metal surface is known to produce a highly concentrated field which is largely confined near the metal surface. This near field effect has long been studied and explored for spectroscopic and optical imaging [1-2]. One of the earliest and well known applications is the dramatic enhancement of Raman spectroscopic signal, as reported by Richard P. Van Duyne in 1977 with the use of Ag electrode for the deposition (adsorption) of the analytes [3]. Interestingly it was found that 10^5-10^6 fold increase of the Raman signal was achieved upon roughening of the Ag surface compared to the signal detected by using a smooth Ag surface. This modified Raman spectroscopy and its continuously improved version has come to be known as the surface-enhanced Raman spectroscopy (SERS) [4]. Meanwhile, further studies on the large enhancement produced by a roughened adsorption surface and colloidal systems [5-7] have led to the idea concerning the role of local hotspots that are mainly responsible for the generation of very strong local near-field and a huge enhancement factor for the Raman signal [8]. This idea was first implemented as tip-enhanced Raman spectroscopy (TERS) using the metal tips of AFM or STM [9-12]. TERS has the advantage of overcoming of the drawbacks of SERS which was shown to have its enhancement depending on the specific nature of the adsorbed analytes. It also has the additional advantage of offering Raman spectroscopy with extremely high spatial resolution [13]. The possibilities of nanoscaled chemical imaging [14-15] and single molecule TERS [16-17] have been reported.

While the estimated enhancement factor of TERS has been formulated in terms of the far-field and near-field intensities of the Raman signals using metallic tips [18-19], little has been reported in the literature about the effects of the size and shape of the tip itself on the spectral and spatial distributions of the near-field enhancement factor around the metal tip. Reported in this paper is the result of a

IOP Conf. Series: Journal of Physics: Conf. Series **1057** (2018) 012008 doi:10.1088/1742-6596/1057/1/012008

computational study on the detailed variations of the spectral profiles and spatial distributions of the near-field enhancement factor around a simple rectangular Ag tips of various sizes and shapes generated by a normal incident electromagnetic wave of certain polarization. For this purpose, the computation of the electric field is performed employing the surface integral equation (SIE) method [20]. Finally, the optimal geometrical parameters yielding the most favorable field enhancement is presented.

2. Computational method

There are many kinds of computational methods for calculating the electric fields scattered from a scatterer. One of these is the integral equations (IE) method which solves electromagnetic scattering problem by representing the scattered field in terms of its associated Green's function [21]. Generally, the integral equations methods are divided into two types, one employs the volume integral equation (VIE) based on the discrete dipole approximation (DDA) and the other one uses surface integral equation (SIE) involving the equivalent electric and magnetic surface current densities. This second method is the one adopted in this study.

The IE method is known to yield reasonably small errors with the requirement of considerably smaller computational memory compared to FDTD and FEM methods which require a large memory for background space discretization. This requirement is avoided in the calculation by means of integral methods, in particularly the surface integral equation method. In addition to that, SIE method can be applied to scatterer of arbitrary shapes and the calculation can be performed on the mathematical surface surrounding the scatterer instead of carrying out the calculations over the entire volume. This method therefore works best for piecewise homogeneous media [22].

Basically, the total electric field in a light scattering problem can be obtained by solving the Helmholtz equation:

$$\nabla \times \nabla \times \mathbf{E}_\gamma(\mathbf{r}) - k_\gamma^2 \mathbf{E}_\gamma(\mathbf{r}) = i\omega\mu_\gamma \mathbf{j}(\mathbf{r}). \quad (1)$$

$\gamma = 1, 2$ correspond to the regions outside and inside the scatterer indicated in figure 1 and $\mathbf{j}(\mathbf{r})$ is the volume current density representing the effect of the scatterer.

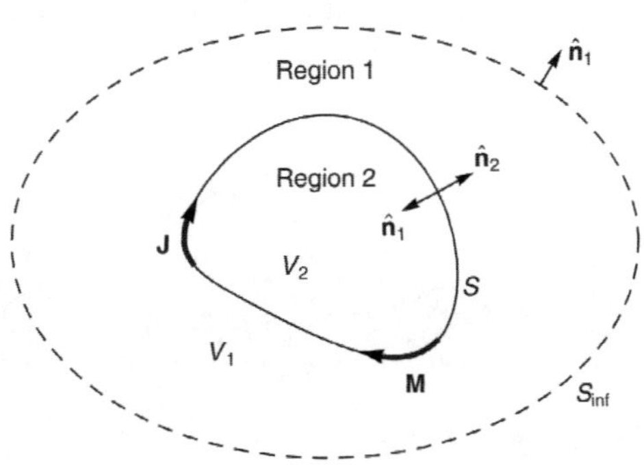

Figure 1. Geometry of the electromagnetic scattering problem [20]

The complete solution of this differential equation consists of the incident electric field, given by:

$$\mathbf{E}_1^{inc}(r') = i\omega\mu_1 \overleftrightarrow{\mathbf{G}}_1(\mathbf{r},\mathbf{r}') \cdot \mathbf{j}(\mathbf{r}')\, dV, \quad (2)$$

and the scattered field which is included in the total electric field as presented by the following equation [20]:

$$\mathbf{E}_\gamma(\mathbf{r}) = \begin{Bmatrix} - \\ + \end{Bmatrix} \oint_{\partial V_\gamma} \left[\frac{\omega\mu_\gamma}{i} \overleftrightarrow{\mathbf{G}}_\gamma(\mathbf{r},\mathbf{r}') \cdot \mathbf{J}_s(\mathbf{r}') - \left[\nabla' \overleftrightarrow{\mathbf{G}}_\gamma(\mathbf{r},\mathbf{r}')\right] \times \mathbf{M}_s(\mathbf{r}') \right] dA' + \begin{cases} \mathbf{E}_\gamma^{inc}(\mathbf{r}), \gamma = 1; r \in V_1 \\ 0, \qquad \gamma = 2; r \in V_2 \end{cases}. \quad (3)$$

In this equation, the dyadic Green's tensor $\overleftrightarrow{\mathbf{G}}_{\gamma}(\mathbf{r},\mathbf{r}')$ is the solution of the corresponding Helmholtz equation with a single point source at $\mathbf{r}=\mathbf{r}'$ as written below.

$$\nabla\times\nabla\times\overleftrightarrow{\mathbf{G}}_{\gamma}(\mathbf{r},\mathbf{r}')-k_{\gamma}^{2}\overleftrightarrow{\mathbf{G}}_{\gamma}(\mathbf{r},\mathbf{r}')=\overleftrightarrow{\mathbf{1}}\delta(\mathbf{r}-\mathbf{r}'). \quad (5)$$

Equation (3) shows that the total electric field can be obtained in either region for the source in the associated region, provided the dyadic Green's tensor $\overleftrightarrow{\mathbf{G}}_{\gamma}(\mathbf{r},\mathbf{r}')$ and the equivalent electric and magnetic surface current densities $\mathbf{J}_{s}(\mathbf{r}')$ and $\mathbf{M}_{s}(\mathbf{r}')$ are known. The dyadic Green's tensor $\overleftrightarrow{\mathbf{G}}_{\gamma}(\mathbf{r},\mathbf{r}')$ can be related with the homogeneous scalar Green's function according to the following expression [23].

$$\overleftrightarrow{\mathbf{G}}_{\gamma}(\mathbf{r},\mathbf{r}')=\left[\overleftrightarrow{\mathbf{1}}+\frac{\nabla\nabla}{k_{\gamma}^{2}}\right]g_{\gamma}(\mathbf{r},\mathbf{r}')=\left[\overleftrightarrow{\mathbf{1}}+\frac{\nabla\nabla}{k_{\gamma}^{2}}\right]\frac{e^{ik_{\gamma}|\mathbf{r}-\mathbf{r}'|}}{4\pi|\mathbf{r}-\mathbf{r}'|}. \quad (5)$$

Meanwhile the equivalent current densities are written as linear combinations of N linearly independent basis functions [24] as

$$\mathbf{J}_{s}(\mathbf{r}')=\sum_{n=1}^{N}\alpha_{n}\,\mathbf{f}_{n}(\mathbf{r}'), \quad (6)$$

and

$$\mathbf{M}_{s}(\mathbf{r}')=\sum_{n=1}^{N}\beta_{n}\,\mathbf{f}_{n}(\mathbf{r}'). \quad (7)$$

where the basis functions $\mathbf{f}_{n}(\mathbf{r}')$ are generally taken to be the Rao-Wilton-Glisson (RWG) basis functions [25] using triangular meshes. The n-th basis function is then defined as:

$$\mathbf{f}_{n}(\mathbf{r}')=\begin{cases}\pm\frac{L_{n}}{2A_{n}^{\pm}}(\mathbf{r}-\boldsymbol{p}_{n}^{\pm}), \mathbf{r}\in T^{\pm}\\ 0 \qquad\qquad\qquad , \mathbf{r}\notin T^{\pm}\end{cases}. \quad (8)$$

In equation (8), L_{n} is a n-th common edge length that separates a pair of adjacent triangular meshes (T_{n}^{+} and T_{n}^{-}). The "+" and "-" notations denote respectively the divergent and convergent surface current directions on the scatterer surface, while $A_{n}^{\pm}$ is the area of mesh, $\mathbf{r}$ denotes the observation point, and $\mathbf{p}_{n}^{\pm}$ are the $A_{n}^{\pm}$ free vertices of the meshes.

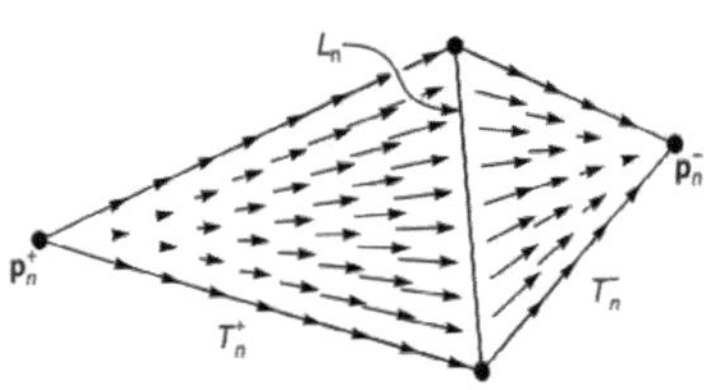

Figure 2. Illustration of the RWG basis functions [25]

Next, the expansion coefficients of the equivalent current densities $\{\alpha_{n}\}$ and $\{\beta_{n}\}$ are determined by substituting Equation (6)-(7) into the electric field and magnetic field integral equations given below.

$$\left(\frac{\omega\mu_{\gamma}}{i}\oint_{S}\overleftrightarrow{\mathbf{G}}_{\gamma}(\mathbf{r},\mathbf{r}')\cdot\mathbf{J}_{s}(\mathbf{r}')\,dA'-\oint_{S}\left[\nabla'\overleftrightarrow{\mathbf{G}}_{\gamma}(\mathbf{r},\mathbf{r}')\right]\times\mathbf{M}_{s}(\mathbf{r}')\,dA'\right)_{\tan}=\begin{cases}\left(\boldsymbol{E}_{\gamma}^{inc}(\mathbf{r})\right)_{\tan}, \gamma=1;\mathbf{r}\in S\\ 0 \qquad\qquad , \gamma=2;\mathbf{r}\in S\end{cases} \quad (9)$$

$$\left(\frac{\omega\varepsilon_{\gamma}}{i}\oint_{S}\overleftrightarrow{\mathbf{G}}_{\gamma}(\mathbf{r},\mathbf{r}')\cdot\mathbf{M}_{s}(\mathbf{r}')\,dA'-\oint_{S}\left[\nabla'\overleftrightarrow{\mathbf{G}}_{\gamma}(\mathbf{r},\mathbf{r}')\right]\times\mathbf{J}_{s}(\mathbf{r}')\,dA'\right)_{\tan}=\begin{cases}\left(\boldsymbol{H}_{\gamma}^{inc}(\mathbf{r})\right)_{\tan}, \gamma=1;\mathbf{r}\in S\\ 0 \qquad\qquad , \gamma=2;\mathbf{r}\in S\end{cases} \quad (10)$$

These equations are solved for $\{\alpha_n\}$ and $\{\beta_n\}$ using the so-called Method of Moments (MoM) [24], where a set of expansion functions and a testing procedure are developed and used to derive the elements of the moment matrices. The resulted $\mathbf{J}_S(\mathbf{r}')$ and $\mathbf{M}_S(\mathbf{r}')$ together with $\overleftrightarrow{\mathbf{G}}_\gamma(\mathbf{r},\mathbf{r}')$ from equation (5) are then substituted into equation (3) to give the total electric field determined by equation [26]. The resulted field $\mathbf{E}_\gamma(\mathbf{r}')$ is further normalized by the incident field $\mathbf{E}_\gamma^{inc}(\mathbf{r}')$ to given the associated enhancement factor.

The geometry of the meshes used in this study are created with the total number ranging within 800-1500 for optimizing the computation time and memory. There is no significant changes of the result for finer meshes. The frequency-dependent permittivity of the silver scatterer is taken from the experimental data of Babar and Weaver [27].

3. Size effect

For the study of size effect on the near field spectra and its spatial distributions, we consider the simple case of a rectangular silver (Ag) nanotape of 5 nm thickness and 5 nm width as shown in figure 3. In the following computation, we restrict ourselves to the case of transversal electromagnetic plane waves incident on the nanotape surface with its electric field polarized along the x-axis direction. The size effect is first investigated by computing the near-field spectra at one end of the tape located at x = (ℓ/2 + 0.1) nm, y = z = 0 nm, where ℓ is the tape length, as well as the spectral variations for various tape lengths.

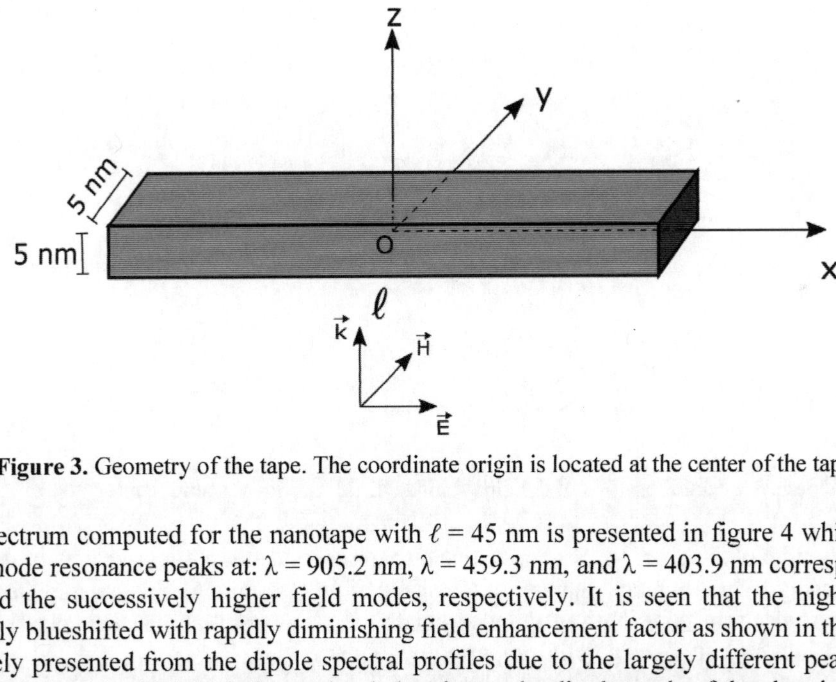

Figure 3. Geometry of the tape. The coordinate origin is located at the center of the tape.

The spectrum computed for the nanotape with ℓ = 45 nm is presented in figure 4 which shows that the multimode resonance peaks at: λ = 905.2 nm, λ = 459.3 nm, and λ = 403.9 nm corresponding to the dipole, and the successively higher field modes, respectively. It is seen that the higher modes are increasingly blueshifted with rapidly diminishing field enhancement factor as shown in the inset, which is separately presented from the dipole spectral profiles due to the largely different peak values. The highest spectral peak which is designated as belonging to the dipole mode of the electric field is found to be the lowest mode found for the tape. The basically similar patterns of the spectra are also found in the cases of different tape lengths. However, the field enhancement factor of the resonance modes are generally found to decrease with shorter tape length. This will be further elaborated figure below.

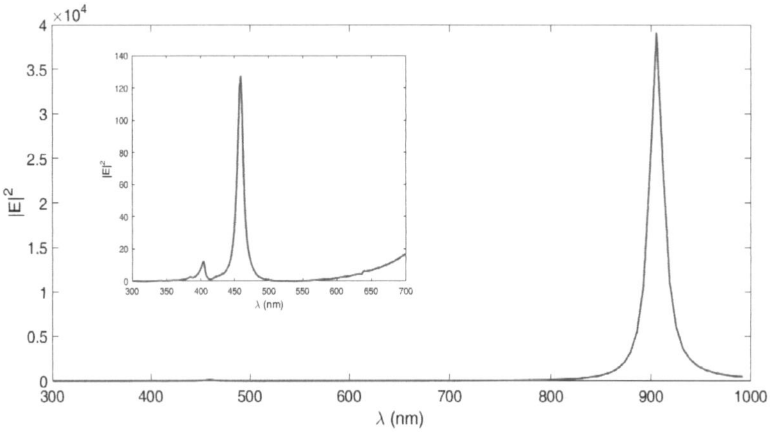

Figure 4. Near field spectrum generated at the end of an Ag nanotape of $\ell = 45$ nm. The inset shows the much lower spectral peaks at $\lambda < 750$ nm

In view of the much more dominant appearance of the dipole mode spectral profile, the following analysis will be focused on the electric dipole spectra calculated for various tape lengths. The result is presented in Fig. 5 for a number of tape lengths. Two features stand out in this figure. Firstly, there is a consistent redshift of the spectral profiles with increasing tape length, which is related with the larger

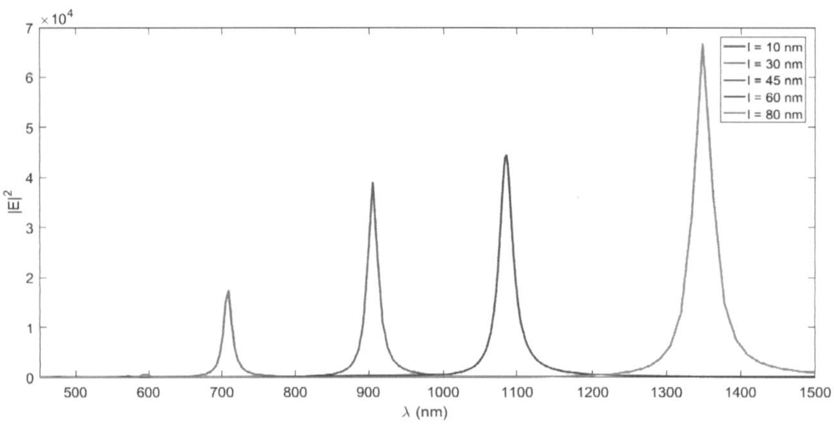

Figure 5. Spectral profiles of the dipole mode $|E|^2$ for various tape lengths (ℓ)

oscillation path length of the electron. Secondly, one also observes a monotonous growth of the spectral peak with increasing ℓ without much changes of the shape of the profiles. Again, this may be due to the larger numbers of electrons contributing to the associated resonant oscillations. This clearly suggests that higher near field intensity is generally supported by longer ℓ.

We turn next to the spatial distributions of the dipole fields at the corresponding peak wavelengths and the variations with respect to changes in ℓ. The computed results are described in figure 6 for several tape lengths indicated in the figure. One sees that the induced electric fields are generally concentrated near the end surfaces of the tape, which decays very fast away from the surface, which is

IOP Conf. Series: Journal of Physics: Conf. Series **1057** (2018) 012008 doi:10.1088/1742-6596/1057/1/012008

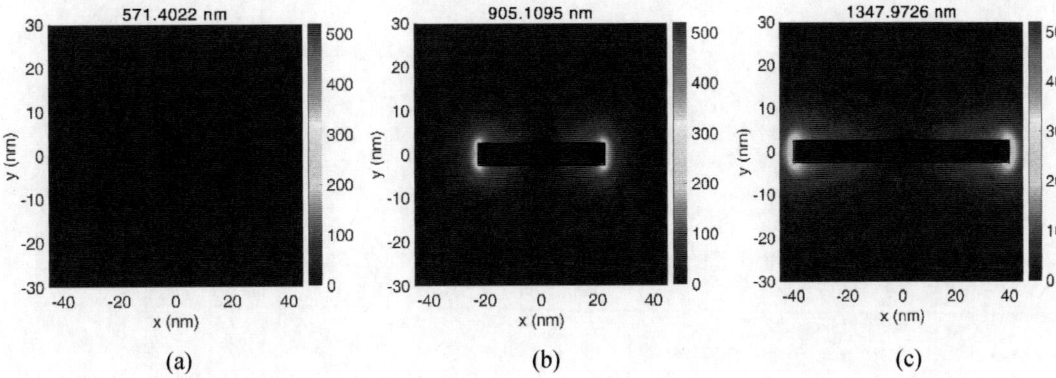

Figure 6. Field distributions around the side surface of the nanotape of various tape lengths, (a) $\ell = 10$ nm, (b) $\ell = 45$ nm, (c) $\ell = 80$ nm, computed for the near fields at the corresponding dipole spectral peaks.

indicative of the typical feature well known for induced surface plasmon field. The frequency redshift is likely related with the longer oscillation path length of the surface electrons. We note further that one barely observes the near field at the tape surface for $\ell = 10$ nm, while a clearly visible larger enhancement with more spread out spatial distribution of the field is seen for $\ell = 80$ nm, suggesting the benefit of longer tape length.

4. Shape effect

For the investigation of the shape effect, we consider the equilateral triangular Ag tape shown in figure 7. The effects are represented by the variations of spectral profile and spatial distribution of optical near field induced at the tape surface for varied apex angle θ as indicated in the figure.

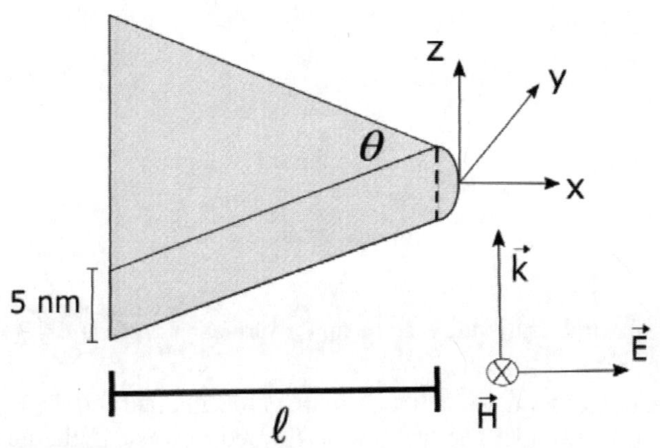

Figure 7. Illustration of the geometry of triangular tip, angle tip (θ) and tip length (ℓ) are varied, and the thickness is fixed at 5 nm. The radius of curvature of the rounded tip is determined as 0.25 nm

The effects of the tape geometry on the near-field at the tip are studied by comparing the dipole field spectral profiles and its spatial distributions computed for various values of the tape parameters (ℓ, θ). To begin with, a comparison is made between the field enhancement generated at the tip of the triangular tape and at the end point of the rectangular, both computed at their corresponding peak wavelengths. The result given in figure 8 for certain ℓ and θ clearly exhibits a larger (more than 4 fold) enhancement effect obtained by the tip geometry.

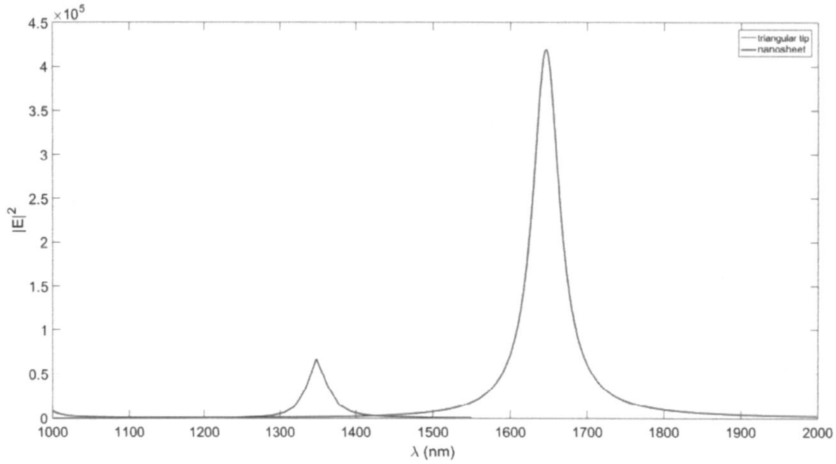

Figure 8. Comparison of spectral profiles induced at the tip of a triangular Ag nanotape of ℓ = 80 nm

In order to explore the optimal geometry of the triangular nanotape for the largest field enhancement at the tip, further variations of ℓ and θ are separately studied. Presented in figure 6 is the result showing the effect of computed for various θ. It is seen that the field enhancement factor varies

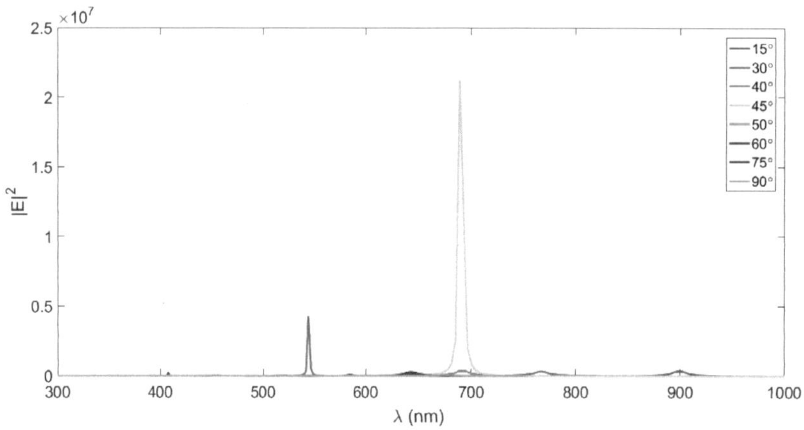

Figure 9. Spectral distiributions at the tip for various tip angle at ℓ = 50 nm

respectively on θ, with the obviously dominant maximum value attained at $\theta = 45°$ recurring at $\lambda = 688.9$ nm. Further exploration of the optimal geometry is performed by varying the tape length (ℓ) with θ = 45°. The result of the computed $|E|^2$ is depicted in figure 10. It is clear that both the peak wavelength of the dipole field as well as the field enhancement factor do not show the monotonous changes with respect to both ℓ and θ. Nevertheless, the field enhancement factor is found to exhibit its largest value 10^7 for θ = 45° and ℓ = 50 nm at $\lambda = 688.9$ nm.

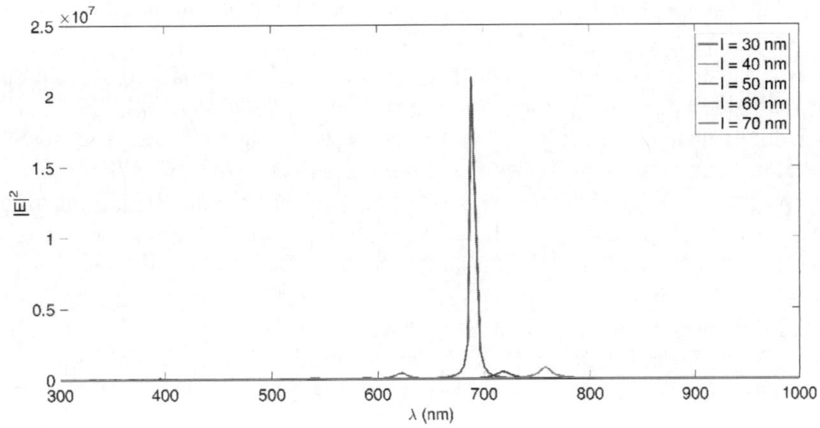

Figure 10. Spectral distiributions of triangular nanotape at optimal angle estimation by varying its tip length at $\theta = 45°$

5. Conclusions

The results of this study have shown the geometrical effects on the near-field enhancement and its spectral shift generated at the tip of a silver nanotriangular tape illuminated by a normal incident plane wave polarized along the long axis of the tape. It is shown that the dipole field generally dominates the spectra generated on the silver nanotape surfaces. Further, consistent monotonous spectral frequency shift and field enhancement factor variations are found in the case of rectangular tape of various tape length. On the other, both the spectral shift and field enhancement factor variations are shown to exhibit complicated non-monotonous variation with the tip angle θ and the length of the triangular Ag nanotape ℓ. It is shown that the maximum enhancement factor of 10^7 is attained for $\ell = 50$ nm, $\theta = 45°$ at resonance wavelength of $\lambda = 688.9$ nm.

References

[1] Maier S A 2004 *Plasmonics : Fundamentals And Applications* (New York: Springer)
[2] Chang R K and Furtak T E 1982 *Surface-Enhanced Raman Spectroscopy* (New York: Plenum)
[3] Jeanmaire D L and Van Duyne R P 1977 Surface raman spectroelectrochemistry: Part I. Heterocyclic, aromatic, and aliphatic amines adsorbed on the anodized silver electrode *J. Electroanal. Chem.* **84** 1-20
[4] Moskovits M 2005 Surface-enhanced Raman spectroscopy: A brief retrospective *J. Raman. Spetrosc.* **36** 485-496
[5] Shalaev V M 2000 Nonlinear optics of random metal-dielectric films *Phys. Rev. B* **158** 13265-288
[6] Grésillon S, Aigouy L, Boccara A C, Rivoal J C, Quelin X, Desmarest C, and Shalaev V M 1999 Experimental observation of localized optical excitations in random metal-dielectric films. *Phys. Rev. Lett.* **82** 4520-23
[7] Markel V A, Shalaev V M, Zhang P, Huynh W, Tay L, Haslett T L, and Moskovits M 1999 Near-field optical spectroscopy of individual surface-plasmon modes in colloid clusters *Phys. Rev. B* **59** 10903-09
[8] Wessel J 1985 Surface-enhanced optical microscopy *J. Opt. Soc. Am. B* **2** 1538-41
[9] Stöckle R M, Suh Y D, Deckert V, and Zenobi R 2000 Nanoscale chemical analysis by tip-enhanced Raman spectroscopy *Chem. Phys. Lett* **318** 131-136
[10] Anderson M S 2000 Locally enhanced Raman spectroscopy with an atomic force microscope *Appl. Phys. Lett* **76** 3130

[11] Hayazawa N, Inouye Y, Sekkat Z, and Kawata S 2000 Metallized tip amplification of near-field Raman scattering *Opt. Commun* **183** 331-336
[12] Pettinger B, Picardi G, Schuster R, and Ertl G 2000 Surface enhanced Raman spectroscopy: Towards single molecular spectroscopy *Electrochemistry (Jpn.)* **68** 942-49
[13] Pettinger B, Schambach P, Villag C J, and Scott N 2012 Tip-Enhanced Raman Spectroscopy: Near-Fields Acting on a Few Molecules *Annu. Rev. Phys. Chem.* **63** 379-99
[14] Hayazawa N, Inouye Y, Sekkat Z, and Kawata S 2002 Near-field Raman imaging of organic molecules by an apertureless metallic probe scanning optical microscope *J. Chem. Phys* **117** 1296–301
[15] Stadler J, Schmid T, and Zenobi R 2010 Nanoscale chemical imaging using top-illumination tip-enhanced Raman spectroscopy *Nano Lett* **10** 4514–20
[16] Domke K F, Zhang D, and Pettinger B 2006 Toward Raman fingerprints of single dye molecules at atomically smooth Au(111) *J. Am. Chem. Soc.* **128** 14721–27
[17] Zhang W, Yeo B S, Schmid T, and Zenobi R 2007 Single molecule tip-enhanced Raman spectroscopy with silver tips *J. Phys. Chem. C* **111** 1733–38
[18] Pettinger B, Picardi G, Schuster R, Ertl G 2004 Nanoscale probing of adsorbed species by tip-enhanced Raman spectroscopy. *Phys. Rev. Lett.* **92** 96101
[19] Pettinger B, Ren B, Picardi G, Schuster R, Ertl G 2005 Tip-enhanced Raman spectroscopy (TERS) of malachite green isothiocyanate at Au(111): bleaching behavior under the influence of high electromagnetic fields. *J. Raman Spectrosc.* **36** 541–50
[20] Kern A M ann Martin O J F 2009 Surface integral formulation for 3D simulations of plasmonic and high permittivity nanostructures *J. Opt. Soc. Am. A* **26** 732-740
[21] Chew W C, Tong M S, and Hu B 2008 *Integral Equation Methods for Electromagnetic and Elastic Waves* (United States of America: Morgan & Claypool)
[22] Gallinet B, Butet J, and Martin O J F 2015 Numerical methods for nanophotonics: standard problems and future challenges *Laser Photonics Rev.* **9** 577-603
[23] Chew W C 1996 *Waves and Fields in Inhomogeneous Media* (New York: IEEE Press)
[24] Gibson W C 2010 *The Method of Moments in Electromagnetics* (Boca Raton: Chapman & Hall/CRC)
[25] Rao S, Wilton D, and Glisson A 1982 Electromagnetic scattering by surfaces of arbitrary shape. *IEEE Trans. Antennas Propag.* **30** 409-418
[26] Raziman T V, Somerville W R C, Martin O J F 2015 Accuracy of surface integral equation matrix elements in plasmonic calculations *J. Opt. Soc. Am. B* **32** 485-492
[27] Babar S and Weaver J H 2015 Optical constants of Cu, Ag, and Au revisited *Appl. Opt.* **54** 477-81

Acknowledgments

This work was supported by Program Unggulan Perguruan Tinggi DIKTI 2017 from the Indonesian Ministry of Research, Technology, and Higher Education, under Contract No. 3091s/I1.C01.2/KU/2017.

IOP Conf. Series: Journal of Physics: Conf. Series **1057** (2018) 012009 doi:10.1088/1742-6596/1057/1/012009

Fabrication of nanostructure grating polymer based coupling element for Surface Plasmon Resonance (SPR) sensors and its spectral reflectance characteristics

Jalu Setiya Pradana*, Siti Chalimah and Rahmat Hidayat#

Physics of Magnetism and Photonics Research Division, Faculty of Mathematics and Natural Sciences, Bandung Institute of Technology, Indonesia

E-mail: * jspradana@gmail.com, # rahmat@fi.itb.ac.id

Abstract. In this paper, we present nanostructure grating polymer, fabricated by employing the nano-imprint lithography technique, with gold nanolayer on its top surface as a coupling element for SPR sensors. The experiment results showed that the diffracted beams for $m = +1$ were suppressed, which indicates the SPR wave excitation in this structure. For further clarification, SPR spectral shape simulation has been also done by using the Rigorous Coupled-Wave Analysis (RCWA) method. The simulation results are in agreement with the observed SPR dips in this experimental work.

1. INTRODUCTION

Surface Plasmon Resonance (SPR) is a resonance phenomenon between electrons oscillation on the metal surface and an incident electromagnetic wave causing quantized plasmon oscillations on metal surfaces. This SPR phenomenon produces an evanescent wave with strong electric field, which is considered can be utilized for enhancing the light excitation or light absorption of a tiny amount of analyte molecules on the metal surface. Therefore, this SPR phenomenon can be used for improving the sensitivity of optical-based bio-chemical or bio-medical sensors. [1]

There are several methods commonly applied to excite SPR wave, i.e. by using a prism, grating, and nanoparticles. Incident light coupling by using a prism, which is well known as the Kretschmann and Otto configurations, are the most commonly implemented method in SPR spectroscopy. However, this method requires delicate measurement set-up of their instrumentation so that rather difficult to be built as a portable or mobile sensor. In contrast, SPR excitation by using a grating as a coupling element requires simpler instrumentation setup. However, the fabrication of this coupling element requires more delicate technique. Moreover, this method also requires more complicated analysis on the measured SPR spectra, particularly when the grating shape is not simple sinusoidal form.

In grating coupling method, SPR excitation occurs when the resonance condition takes place where the propagation constant of the incident along the grating surface k_x and the surface plasmon wave vector k_{sp} is given by the following relationship

$$k_{mx} = \frac{n_{inc} \sin \theta_{inc}\, 2\pi}{\lambda_0} - m\frac{2\pi}{\Lambda_x} = k_{sp} \tag{1}$$

where n_{inc} is dielectric refractive index, θ_{inc} is incident angle, λ_0 is light wavelength in vacuum, m is diffraction order, and Λ_x is grating periodicity. In the present work, we have successfully fabricated gratings having surface corrugated structure with square wave pattern and sub-micron periodicity.

These grating structures were made from hybrid polymer by simple nano-imprint lithography method, which then covered by a very thin gold layer of around 50 nm in thickness. The advantages of hybrid polymer are able to be shaped from its precursor gel by either photo- or thermal-polymerization, resulting a structure with good mechanical strength, thermal stability and optical clarity. [2]

In order to get an insight on the possible reflectance spectra or SPR spectra in those structures, calculations based on the Rigorous Coupled-Wave Analysis (RCWA) method have been also carried out. The RCWA method is different from the coupled wave analysis (CWA) because RCWA is a fully vectorial calculation for solving the Maxwell equation without using any approximation as in CWA. [3] In RCWA method, the light waves and the medium structure are formulated into their Fourier series expansions. This approach suits well with the grating structure in the present work. The calculation results are then a representation of electric fields in each mode. Moreover, as another advantage of RCWA, the differential equations of Maxwell's equation are transformed into matrix forms. The wave equation (for two-dimensional structure) then can be written as

$$\frac{\partial^2}{\partial z^2}\begin{bmatrix} s_x \\ s_y \end{bmatrix} - \Omega^2 \begin{bmatrix} s_x \\ s_y \end{bmatrix} = 0 \qquad (2)$$

where s_x and s_y are the electric fields which can be found from the eigenvector solution of Eq. (2) and Ω stands for wave propagation constants. All s_x, s_y and Ω are in matrix form. [3]

2. EXPERIMENTS

2.1. Precursor gel preparation

The precursor gel was prepared by the sol-gel method from 3-(Trimethoxysilyl) propyl methacrylate 98% (TMSPMA, Aldrich) as in our previous work. TMSPMA was previously dissolved in ethanol and added by HCl 1 N as the catalyst. The precursor solution was continuously stirred 260 rpm for 72 hours at 50°C. The result of this process was a gel polymer precursor, which was then added by Irgacure-819 (Ciba Specialty Chemical Inc.) photo-initiator. This process was done in dark room to avoid undesired photo-polymerization.

2.2. Grating SPR fabrication

The sub-micron polymer grating structure was then made from that precursor gel using the nano-imprint lithography method. The precursor was poured over a silicon stamp, followed by a vacuum process for 30 minutes. The next step was UV-curing process where that precursor gel was irradiated by 405 nm laser light for 15 minutes. Finally, the formed grating structure was coated with the gold layer of around 50 nm in thickness by the sputtering technique.

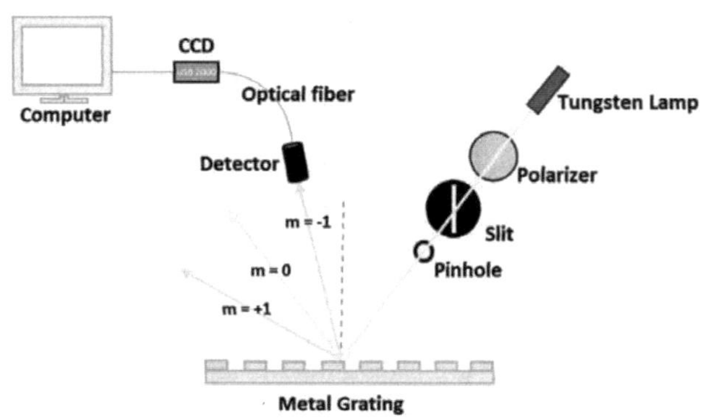

Figure 1. Schematic diagram of the measurement setup of SPR spectrum.

IOP Conf. Series: Journal of Physics: Conf. Series **1057** (2018) 012009 doi:10.1088/1742-6596/1057/1/012009

2.3. SPR Spectra Characterization

The fabricated grating SPR elements were characterized by measuring the reflectance spectra with a measurement setup as shown in Fig. 1. A tungsten lamp was used as the polychromatic light source. The incident light was polarized either to be the *p*- or *s*- (TM or TE) polarization light beam by using a polarizer. A slit and a pinhole were used to keep the light beam with an appropriate small diameter. The reflected light beam was captured by a CCD based spectrophotometer that was connected to a computer.

2.4. Simulation

The RCWA calculations were performed by using a modified computation program initially developed by the Zhang group of the Georgia Institute of Technology. [4] The grating structure model in the simulation was made to be same as the fabricated grating structure.

3. RESULTS AND DISCUSSION

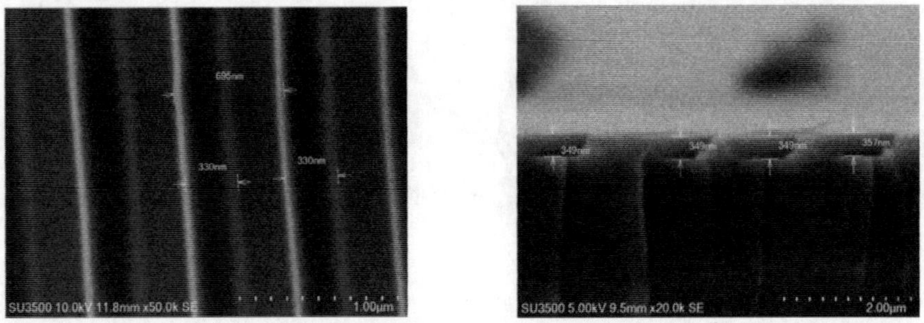

(a) (b)

Figure 2. The SEM images obtained from (a) the surface and (b) cross-section of the fabricated polymer grating structure with grating periodicity of about 700 nm.

Figure 2 shows the scanning electron microscopy (SEM) images indicating the formation of surface corrugated or surface relief grating structure with a periodicity of about 700 nm. This formed grating structure has almost the same structure and periodicity as the template. Grating structure with periodicity up to around 150 nm can be made by using this polymer precursor gel and nano-imprint lithography.

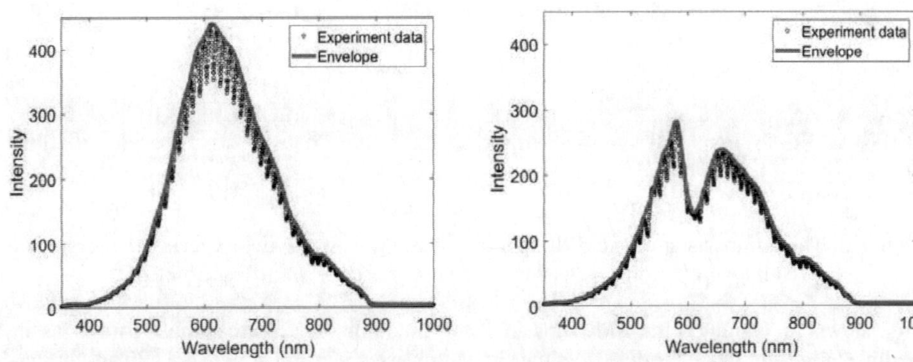

(a) (b)

Figure 3. The reflection spectra obtained from the diffracted light beams at diffraction order $m = +1$ and incident angle 45° for (a) the s-polarization (TE mode) and (b) the p-polarization (TM mode).

The diffraction light beams were observed for $m = 0$ and $m = +1$ at incident angles of 40°, 45°, and 50°. The zero-order diffracted light ($m = 0$) looks like a polychromatic beam spot, whereas the first order diffracted light ($m = +1$) looks like a monochromatic light beam with wavelength depending on the diffraction angle. In order to analyze the complete reflection spectrum at this first diffraction order, we added these first order diffraction spectra for all diffraction angles and took the envelope from the resulting total spectrum, as shown in Figure 3. This envelope spectrum was then divided by the incident light spectrum, taken from a flat Au layer at the same incident angle, to get the complete reflectance spectrum. Figure 4 shows these complete spectra for the s-polarization (TE mode) and the p-polarization (TM mode), giving an evidence of the appearance of SPR dip in the p-polarized light spectrum. Moreover, this figure also shows that this SPR dip was red-shifted with increasing incident angle.

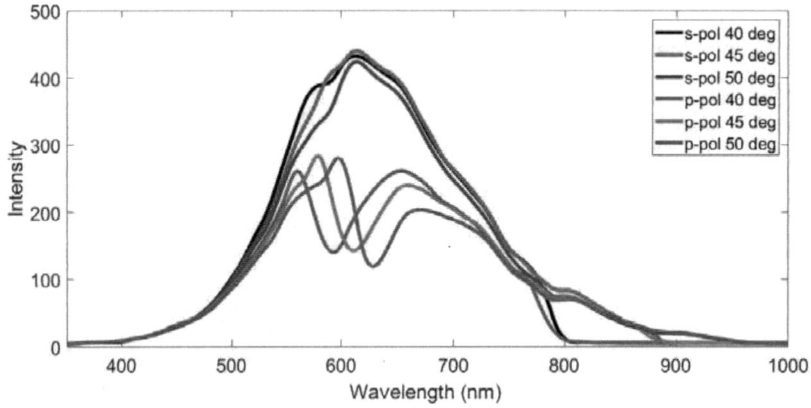

Figure 4. The comparison of the complete spectra of the s-polarization (TE mode) and (b) the p-polarization (TM mode) from the first order (m=+1) diffraction light.

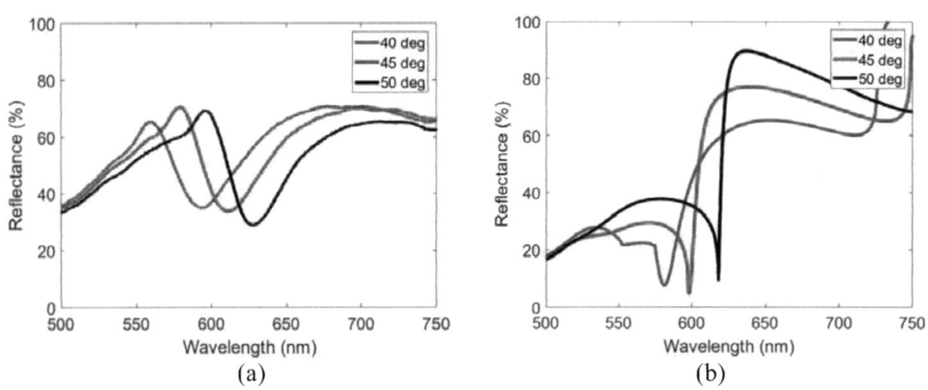

Figure 5. The comparison of the SPR spectra obtained from (a) the experimental results and (b) the simulation results for the first order (m=+1) diffraction light.

As shown in Figure 5, the SPR dips in the simulation spectra are almost similar as in the experimental results. The SPR dips from the simulation spectra are 582, 599, and 620 nm for incident angles of 40°, 45°, and 50°, respectively. The corresponding SPR dips from experimental results are 594, 611, and 627 nm.

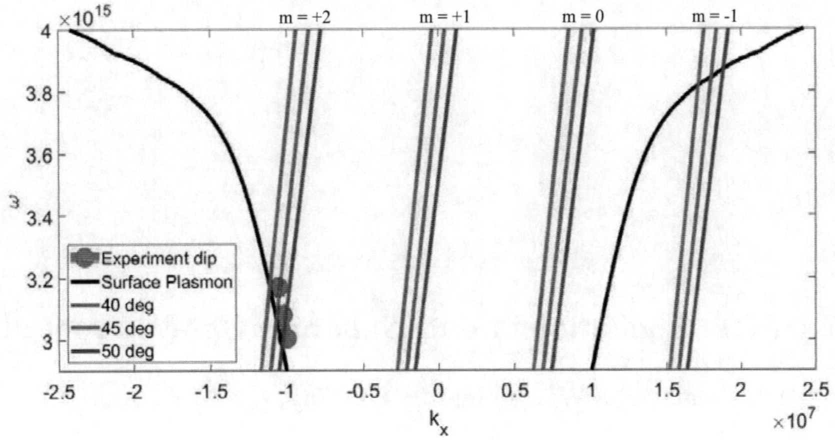

Figure 6. Dispersion relation curves.

In order to clarify the origin of SPR dips, the experimental SPR dips were plotted in the dispersion relation given by Eq. (1), as shown in Fig. 6. Interestingly, the SPR dips are around the blue line for the surface plasmon dispersion relation with $m = +2$. so diffraction order which was generated SPR in the visible light is $m = +2$. The origin of SPR dips seems to be caused by SPR excitation from the second order diffraction, not from the first order diffraction. This fact means that the excitation SPR wave is caused by the second order diffraction light beam and absorbs the light energy at the resonance wavelengths. Due to this light absorption, the first order diffracted light intensity then becomes much weaker.

4. CONCLUSION

We have successfully fabricated some surface-corrugated polymer gratings with sub-micron periodicity by using nano-imprint lithography from hybrid polymer precursor, which can exhibit SPR excitation after the deposition of 50 nm thin gold layer. The SPR dips were observed in the first order (m = +1) diffraction light beams. The observed SPR spectrum shows an agreement with the SPR spectrum from the simulation results based on RCWA method. Interestingly, the dispersion curve shows that SPR dips are from the second order ($m = +2$) diffraction. These facts then lead to a conclusion that, in the present case, the observed SPR in the present case is not directly from the resonance condition with the first order diffraction, but from the second order that has resonance wavelength coincidentally matched with the first order diffraction wavelength.

References

1. Homola, J, (2006), Surface Plasmon Resonance Based on Sensor. Springer Series on Sensor and Biosensor n. 04. Heidelberg GmbH and Co. K, Springer-Verlag Berlin.M. P. Brown and K. Austin, Appl. Phys. Letters **85**, 2503–2504 (2004).
2. Hidayat R *et al* 2012 *Polymer. Adv. Tech.* **23** 1264-1270
3. Peng, S and Morris G.M., Efficient implementation of rigorous coupled-wave analysis for surface-relief gratings, J. Opt. Soc. Am. A 12(5) 1087-1096 (1995)
4. http://zhang-nano.gatech.edu/

Acknowledgment

Authors acknowledge the support from Desentralisasi research program from ITB-Ristekdikti.

The Effect of CTAB addition in ZnO Nanorods Optical Properties

A H Ramelan[1], S Wahyuningsih[1], FN Aini[1], L N M Z Saputri[1]

[1] Inorganic Materials Research Group, Universitas Sebelas Maret, Indonesia
Email : aramelan@mipa.uns.ac.id

Abstract
ZnO nanorods have been successfully synthesized using precursor of zinc nitrate tetrahydrate and *cetyltrimethylammonium bromide* (CTAB) as the surfactant. The composition adopted is specified by the ratio of $Zn(NO_3)_2{\cdot}4H_2O$:CTAB = 1:0, 1:1, 1:2, and 1:3 (w/w) using hydrothermal method at temperature of 180 °C for 24 hours. The resulted ZnO nanorods were characterized by X-Ray Diffraction (XRD), Transmission electron microscopy (TEM) and UV-Vis Spectrophotometer. The presence of CTAB addition was found to increase the crystal size. The TEM characterization results showed the formation of homogeneous rods in nano scale. Modified ZnO nanorods improved optical properties, attaining a band gap energy of ~2.9 eV. The transmittance increases and the reflectance decreases with increasing CTAB.

1. Introduction

ZnO material is a semiconductor material that has high transmitance and reflection[1] that was developed as Anti-Reflection Coating (ARC) material. The developed one-dimensional nanostructures such as nanorod improved optical, electronic to that bulk structured materials.The properties of zinc oxide are strongly depend on its structure, including the morphology, aspect ratio, size, orientation and density of crystal. Moreover, these structural characteristics play an important role in many applications. Therefore, development of a shape controlled zinc oxide synthesis method is indispensable for exploring the potential of this material as a smart and functional material [2]. Though many methods have been proposed to synthesize 1D structure. Recently, large-scale and low-cost synthesis of ZnO nanomaterials with 1-D structure, such as nanorods, have been achieved using a hydrothermal method [3,4]. This brings a good chance to study its optical properties.

CTAB is a cationic surfactant. When CTAB was dissolved in water or ethanol, it will be ionized into CTA^+ and Br–. ZnO, as a polar crystal, has a polar axis and possesses a positive face and a negative face on the crystal due to the asymmetrical distribution of Zn and O atoms along its polar axis, and the positive face (0 0 0 1) was occupied by Zn atoms while the negative face (0 0 0 1) was distributed by O atoms. So CTA^+ and Br^- will affect the crystal size and morphology of ZnO by electrostatic attraction[5], where crystal size dan morphology will affect in ZnO optical properties. In this paper, the effects of CTAB on the crystal size and morphologies of hydrothermal synthesis of ZnO nanorods were investigated.

2. Materials And Sample Preparation

2.1 Materials

Zinc nitrate tetrahidrate ($Zn(NO_3)_2.4H_2O$) Aldrich) and ammonium hydroxide (NH_4OH, Aldrich), deionized water were used as the starting materials. N-cetyl-N,N,N-trimethyl ammonium bromide (CTAB) was used as the surfactant.

2.2 Synthesis of ZnO nanorods

ZnO nanorods were synthesized by hydrothermal method. 0,1M $Zn(NO_3)_2.4H_2O$ was dissolved in 100 mL of deionized water and stirred magnetically for about an hour at room temperature. Ammonium hydroxide was added into solution until pH of the solution was maintained as 7, followed by virgous stirring for an hour. A white precipitate ($ZN(OH)_2$ was produced which was collected by centrifugation and washed throughly with deionized water. Then the precipitate was transferred into 100 mL water containing CTAB. The stoichiometric ratio of $Zn(NO_3)_2.4H_2O$ and CTAB was taken as 1:0; 1:1; 1:2; and 1:3 (w/w). The solution was transferred into a Teflon-line coated and then heated in a electrical oven at a constant temperature of 180 °C for 24 h. After completion of the reaction, it was cooled to room temperature and powder sample was collected by centrifugation. Powder sample was thoroughly washed with deionized water and ethanol. Finally, sample was dried at 80 °C for 12 h.

2.3 Sample Characterization

The crystal structure of teh powder was studied by powder X-ray diffractometer using monochromatic nickel filtered CuKα (λ=1.5416 Å) radiation. The particles morphology was investigated by transmission electron microscopy (TEM-JEM-2010, JEOL). The optical absorption spectrum of the sample was taken in the range between 200-800 nm usig double beam UV-Vis spectrophotometer.

3. Results And Discussion

During hydrothermal process following chemical reactions may have been involved [6].

$Zn^{2+} + 2OH^- \rightarrow Zn(OH)_2$........(1)

$Zn(OH)_2 \leftrightarrow ZnO + H_2O$........(2)

$Zn(OH)_2 + 2OH^- \leftrightarrow [Zn(OH)_4]^{2-}$........(3)

$CTAB \leftrightarrow CTA^+ + Br^-$........(4)

$[Zn(OH)_4]^{2-} + CTA^+ \leftrightarrow CTA^+ - [Zn(OH)_4]^{2-}$........(5)

$CTA^+ - [Zn(OH)_4]^{2-} \leftrightarrow ZnO + H_2O + CTA^+$........(6)

Sun et al [4] has suggested that CTAB not only accelerate the reaction of the growth units but also leads to their oriented growth due to the presence of CTAB reduce the surface tension of the solution which is decreasing the endothermic energy to form a new phase.

The diffractogram of ZnO nanorods shown in Figure 1, the XRD peak position shows that ZnO hexagonal wurzite with lattice constants a = 3.253Å and C = 5.215Å corresponding to the ICSD (No. 67848). Dominant peak is present at $2\theta = 36.52_{(101)}$. The increasing of CTAB addition performed sharp and clear peaks indicating the formation of materials with high crystallinity and intensity that indicated the addition of CTAB enhanced the alignment of nanorods [7].

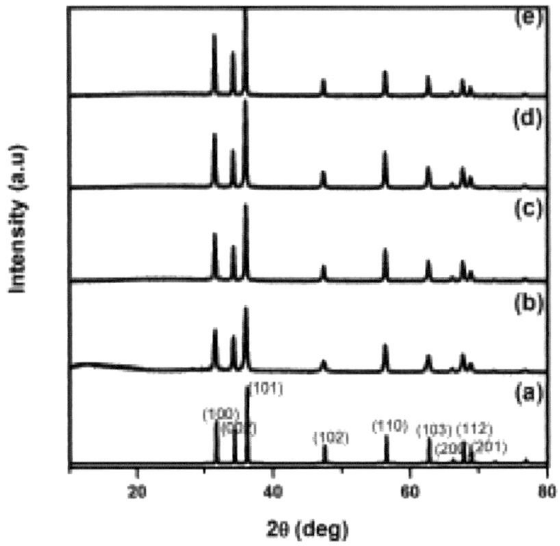

Figure 1. ZnO diffractogram pattern with variations of addition CTAB surfactant (a) ICSD Standard No.67848; (b) ZnO without addition of surfactant; (c) 1: 1 surfactant addition; (d)1:2 surfactant addition; (e) 1: 3 surfactant addition

Crystallite size for ZnO nanorods was calculated using Debye-Scherer formula given in Eq.1 [8]. Increased calcination temperature has an impact on increasing crystallite size.

$$D = \frac{k\lambda}{B\cos\theta} \qquad (1)$$

Where B is full width at half maximum (FWHM), k is constant of proportionality, λ is the X-ray wavelength and θ is the Bragg's diffraction angle. Crystal size of ZnO material shown in Table 1.

Table 1. Average crystal size of ZnO nanorods

Precursor $Zn(NO_3)_2{\cdot}4H_2O$:CTAB (w/w)	B (deg)	2θ (deg)	Cos θ	D (nm)
1:0	0,355	36,610	0,949	23,632
1:1	0,278	36,012	0,951	30,061
1:2	0,272	35,970	0,951	30,755
1:3	0,236	36,010	0,951	35,387

The more CTAB added, the larger crystal size was formed. This is because the more CTAB that added would encourage the nanorods materials formation. TEM imaging results of ZnO nanorods are shown in Fig. 2, indicating that rods have been formed with a tapered end or cone.

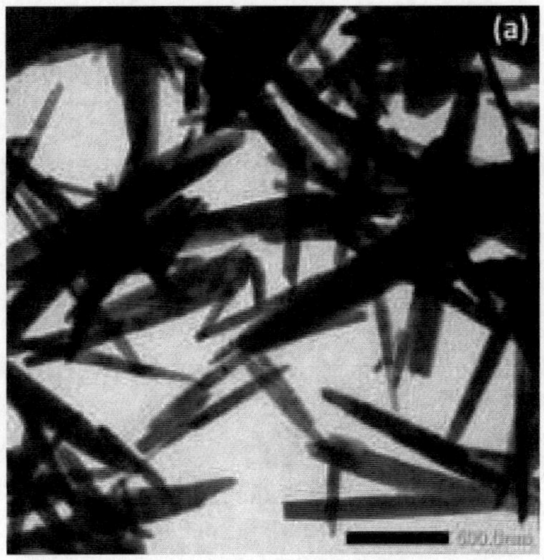

Figure 2. TEM morfology derived from CTAB precursor synthesis, 1:3 (w/w)

The optical studies are performed to evaluate the potentially useful optical quality of the nanorods. UV-Vis absorption spectrum of the ZnO nanorods prepared under hydrothermal conditions is shown in Fig.3. ZnO nanorods absorption peak ranges from 310-330 nm that indicated the material properties was transparent. This properties has advantages such as the optical properties will be increased due to a lot of light was absorbed. The band gap of ZnO nanorods in rasio 1:3 (w/w) is about ~2.9 eV. The increasing crystal size results in less bandgap energy due to quantum confinement effect phenomenon in nanomaterials [9]. This decreasing of bandgap energy becomes an indication of the increase in crystallinity caused by the gathering of ZnO nuclei to form a cluster and eventually become a solid and regular crystallite grain of larger size [10].

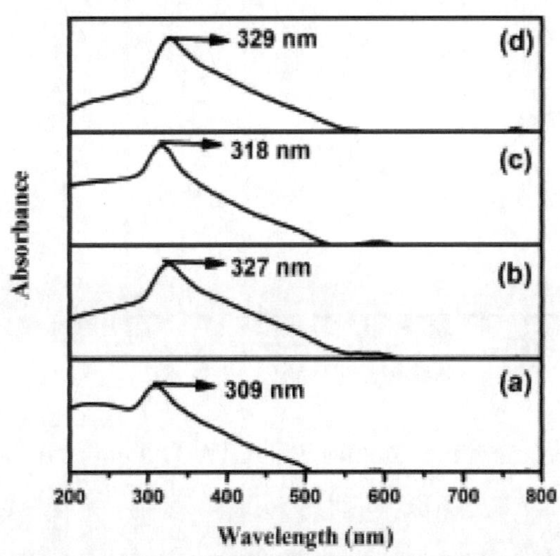

Figure 3. Spectrum absorbance of ZnO nanorods in variations of CTAB surfactant addition (a) ICSD Standard No.67848; (b) ZnO without surfactant addition ; (c) 1:1; (d) 1:2 (e) 1: 3 surfactant addition (w/w)

There is a increase of % T of ZnO nanorods material with increasing of CTAB due to the increasing number of CTAB that caused ZnO nanorods was distributed evenly to increase optical scattering (Figure 4).

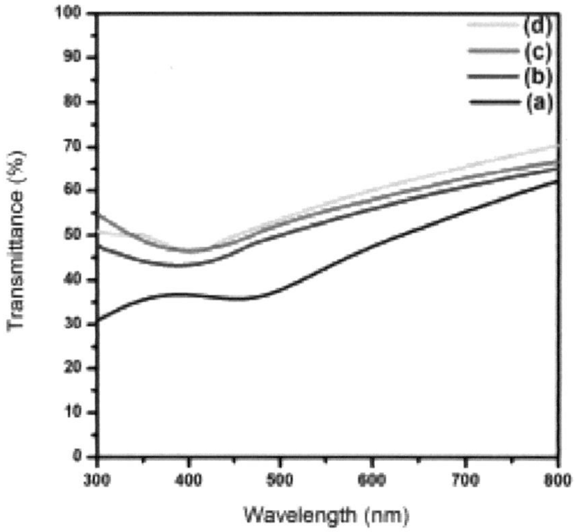

Figure 4. Transmittance spectra of the samples with CTAB additions of (a) 1:0; (b) 1:1; (c) 1:2; (d) 1:3 (w/w)

There is a decrease in% R of material ZnO with increasing of CTAB due to the increasing number of CTAB that caused ZnO nanorods was distributed evenly to increase optical scattering (Figure 5).

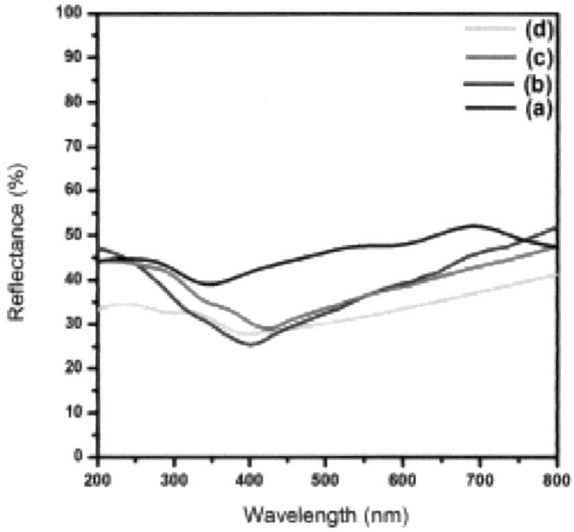

Figure 5. Reflectance spectra of the samples with CTAB additions of (a) 1:0; (b) 1:1; (c) 1:2; (d) 1:3 (w/w)

Relatively larger crystal size tends to decrease the reflectance. This is due to the larger absorption of photons following the Lambert-Beer's law. Smaller size, has shorter photon's internal path and hence more effective surface reflection [11].

4. Conclusion

ZnO nanorods was synthesized by a cost-effective hydrothermal route with addition of CTAB surfactant. The diffraction peak (101) was found stronger and narrower than the other peaks, which indicates a preferential orientation along the c-axis. The addition of CTAB caused increasing crystal

IOP Conf. Series: Journal of Physics: Conf. Series **1057** (2018) 012010 doi:10.1088/1742-6596/1057/1/012010

size that leads to nanorods. Absorbance of ZnO nanorods material indicated that the material absorbs in the UV region. The large crystal size caused a decrease in reflectance, so ZnO nanorod can applied as an anti-reflection (ARC).

Acknowledgments
The authors would like to acknowledge the International Research Colaboration and Scientific Publications Program, Ministry of Research, Technology and Higher Education, Republic of Indonesia, for supporting this research and to the Integrated Mathematics and Natural Science Laboratory of Sebelas Maret University for supporting and providing the facilities for this research.

References

[1] M.H. Aslan, A.Y. Oral, E. Menşur, A.Gül and E.Başaran. 2004. *Sol. Energy Mater Sol. Cells.* **82**(4) 543-552
[2] S. Cho, S-H Jung and K-H. Lee 2008 *J. Phys. Chem.* **112** 12769-12776
[3] N. Yamazoe, G. Sakai, K. Shimanoe 2003 *Catal. Surveys Asia* **1** 63–75
[4] X.M. Sun, X. Chen, Z.X. Deng, Y.D. Li, A 2002 *Mater. Chem. Phys.* **78** 99–104
[5] Y-X. Wang, J. Sun, X. Y. Fan and X. Yu 2011 *Ceram. Int.* **37** 3431-3436
[6] S.D. Shinde, G.E.Patil, D.D.Kajale, D.V.Ahire, V.B.Gaikward and G.H. Jain. 2012 *Int. J. Smart Sensing Intell. Sys.* **5**(1) 57-70
[7] U.N.Maiti, S.Nandy, S.Karan, B. Malik and K.K.Chattopadhyay 2005 *Appl. Surf. Sci.* **254,** 7266-7271(2005).
[8] A. Gupta, H.S. Bhatti, D. Kumar, N.K.Vema, and R.P.Tandon 2006 *Digest Int. J. Nanomat Biostruct* **1** 1-9
[9] G.Pellegrini, G.Mattei and P.Mazzoldi 2005 *J. Appl. Phys.* **97**(7) 6-8
[10] A.HYuwono, B.Liu, J.Xue, J.Wang, H.I.Elim, W.Ji and T.J.White 2004 *J. Mat. Chem.* **14**(20) 2978-2987
[11] B. T. Jheng, P.T Liu, M.C.Wang and M.C.Wu 2013 *Appl. Phys. Lett.* **103**(5) 052904

IOP Conf. Series: Journal of Physics: Conf. Series **1057** (2018) 012011 doi:10.1088/1742-6596/1057/1/012011

Fluorescence spectroscopy characteristics of Indonesian citrus

T H Siregar[1] , U Ahmad[2], Sutrisno[2], and A Maddu [3]

[1]Graduate School of Bogor Agricultural University, Bogor, Indonesia.
[2]Dept. of Mechanical and Biosystem Engineering., Fac. of Agricultural Engineering and Technology, Bogor Agricultural University, Bogor, Indonesia.
[3]Dept. of Physics., Fac. of Natural Sciences and Mathematic, Bogor Agricultural University, Bogor, Indonesia.
E-mail: tikahafzara@gmail.com, uahmad2010@gmail.com, kensutrisno@yahoo.com, akhiruddin@apps.ipb.ac.id

Abstract. Citrus suffering physical damage to the peel would spread polymethoxylated flavone that can be detected by its spectroscopic characteristics. Fluorescence spectroscopy is expected to offer an important means for physical damage detection of Indonesian citrus. This study was conducted to determine the characteristics of fluorescence spectra such as absorbance and fluorescence spectra of four varieties of citrus from Indonesia and identified the fluorescence substance in citrus peels. From this study, the absorbance and fluorescence wavelengths of Indonesian citrus were found to vary from 315 nm to 340 nm and from 500 nm to 555 nm respectively. The fluorescence substances were identified as polymethoxylated flavone such as hesperidin, tangeretin, and nobiletin. The fluorescence spectroscopy could be used to detect defects in citrus peel.

Keywords: *absorbance spectra, citrus, fluorescence spectra*

1. Introduction

A study of spectroscopy characteristics is important to understand the material. Spectroscopy characteristic from fruit substances can be used to build a non-destructive evaluation system. Fluorescence spectroscopy has been used for oil classification [10,12], studying the chemical compounds [5,13], detecting fruit defects such as freeze damage and mold [11,3], detecting the disease stress in citrus plant [2], and quality evaluation of meat product [8,9]; also, fluorescence spectroscopy could be used to evaluate the quality of citrus fruit. Polymethoxylated flavone contained in citrus fruit was one of the fluorescence substances that glowed under UV light excitation. It is possible to detect defects in citrus peel using fluorescence spectroscopy.

Indonesian citrus is an exported commodity [1]. The citrus is subject to damage during handling and transportation. These damaged fruits should be separated as soon as possible to prevent spreading rot starting from the damaged ones. Separating them manually was difficult especially at early stage, because this citrus did not show any visual change before the damaged part changed its color into brown as a browning reaction happened. Fluorescence spectroscopy could be one non-destructive method to detect the damage in citrus peel. The fluorescent substance can facilitate a different approach for early detection of the physically damaged citrus.

Many studies of fluorescence spectra from citrus fruit has been studied. The fluorescence substances in Mandarin citrus had been identified [4]. An investigation of wavelength excitation for fluorescence emission from orange peel exposed to UV light has been done [7]. From this research, it has been known to fluorescence wavelengths of 15 varieties of citrus grown in Japan were 490-540 nm. Fluorescence wave was derived from the substance available in citrus peel. To build up the image processing engine that can detect damage to the Indonesia citrus, data of citrus on fluorescence spectroscopy characteristics is required. Information of fluorescence and excitation wavelengths can be used to build the model to detect the defects of citrus fruit. The purposes of this research were to study the characteristic of fluorescence spectroscopy from Indonesian citrus and identify the fluorescent substance to support the development of a physical damage detection method of Indonesian citrus based on fluorescence spectroscopy.

2. Materials and methods

2.1. Sample preparation

This experiment was carried out with four varieties of citrus. Batu, Medan, Pulung, and Pontianak citrus were the samples as shown in Fig 1. Citrus was harvested from three different islands from Indonesia. These samples were the famous and favorite citrus in Indonesia. The samples were stored in a room with air conditioning set to 25°C for one day. Some peel was taken from the samples for extraction. The procedure to prepare peel extract was as follows; citrus peel by weight of 3.1 g was mashed and mixed with 10 mL of pure n-hexane for spectroscopy as solvent. The extract was stirred and kept for one night at the same temperature as the fruits stored. Citrus peel extract was later used for data retrieval using spectrophotometer for absorbance and fluorescence wavelengths observation. First, the absorbance spectra were recorded. From spectra curve, peak of absorbance spectra were taken as excitation wavelength for measuring itself. After determining the best excitation wavelength, the fluorescence spectra were measured.

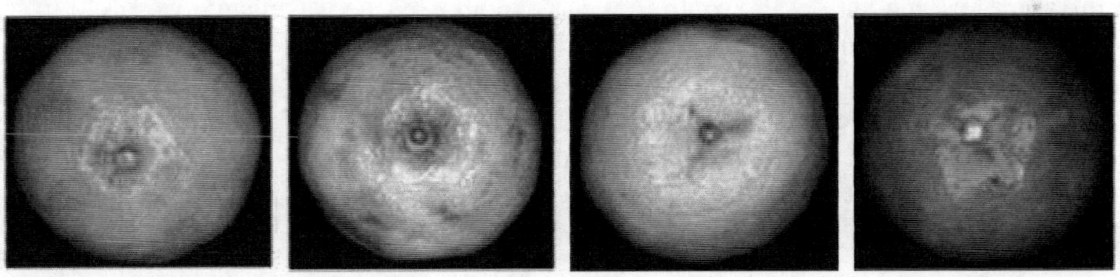

(a) (b) (c) (d)

Figure 1. Samples of Indonesian citrus; (a) Baru, (b) Medan, (c) Pulung, (d) Pontianak

2.2. Spectroscopic measurement

Absorbance spectra was recorded on a UV-Vis Spectrophotometer (Ocean Optics, USA) with spectra bandwidth of 0,2 nm. Fluorescence spectra was recorded with a USB Fluorescence spectrophotometer (Ocean Optics, USA). Excitation light was provided by a UVGL-58 UV lamp with wavelength at 365 nm. Citrus peel extract was placed in quartz cell and spectra was determined.

2.3. Fluorescence substance identification

The fluorescence substance identification was done by testing the Indonesian citrus peel extract with standard solution of Polymethoxylated flavone (Kanglu Biotech) containing hesperidin, tangeretin, and nobiletin using the High Pressure Liquid Chromatografy (HPLC) fingerprint method. This analysis aimed to determine the chemical compound of fluorescence substance contained in citrus peel. For this analysis, citrus peel extract has been made. Indonesian citrus peel was dried with an oven at 45°C for 48 hours. 10 mg of mashed citrus peel was then dissolved with 100 mL of methanol and sonification for 20 minutes. The citrus peel solution was then filtered and the data was recorded using HPLC. HPLC used reverse phase colom (Zorbax sb C-18), solvent gradient system (water / acetonitrile, v / v) used in the mobile phase. The solvent gradient system was listed in Table 1. Then, the retention time from citrus peel solution and standard solution were measured. First, the standard solution and citrus peel solution (10μL each) were separately injected into the HPLC system and then the chromatograms recorded. Retention time of standard solution in chromatograms and the citrus peel solution in the chromatogram were measured. Finally, the hesperidin, nobiletin and tangeritin peaks in the chromatogram of citrus peel solution were identified by comparing the retention time with the standard solution's chromatogram.

Table 1. Chromatographic system conditions

Time (min)	Water (%,v/v)	Acetonitrile (%,v/v)	Elution
0	85	15	Linier gradient
10	60	40	Linier gradient
30	45	55	Linier gradient
40	85	15	Linier gradient

3. Results and discussion

3.1. Absorbance Spectra of Indonesian citrus

The primary peak of absorbance wavelengths from four citruses is almost the same. Fig 2. shows the absorbance spectra of Indonesian citrus. The primary peak appeared in 300 nm to 365 nm. The absorbance peak occurred in the UV region from 315 nm until 340 and then decreased as the wavelength become longer and the second smaller peak was found in 420 nm. From this information, the fluorescence emission would occur when excitated by UV light. To measure the fluorescence spectra for each variety, maximum intensity of absorbance wavelength was used to determine the excitation wavelength. Most of these samples showed that the maximum value of absorbance was in 330 nm until 340 nm. In this study the proper lamp to excitate the sample was 365 nm. It was reported that the excitation wavelength of 365 nm provided better fluorescence spectra of eight varieties Japanese citrus [6]. The second peak appearing in 420 nm to 450 nm referred to another fluorescence substance. It meant that this citrus peel extract contained two kinds of fluorescence substances.

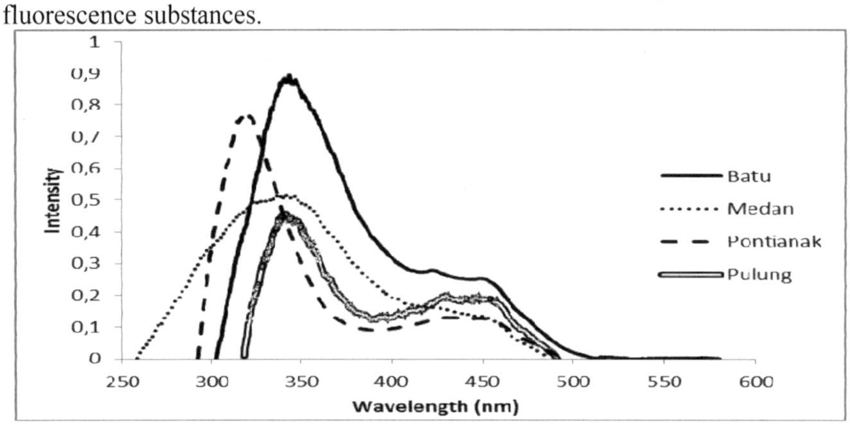

Figure 2. Absorbance spectra of Indonesian citrus

3.2. Characteristic of fluorescence spectroscopy from Indonesian citrus

There were two primary peaks in fluorescence spectra. Fig 3. shows the fluorescence spectra of Indonesian citrus. This fluorescence spectra excitated by 365 nm. Fluorescence spectra occurred in the visible region. The first fluorescence spectra peak occurred from 500 nm until 555 nm and the second occurred from 630 nm until 800 nm. This fluorescence spectra were almost the same with Japanese and Mandarin citrus. The fluorescence spectra peak from fifteen varieties of Japanese citrus appeared in 490 nm to 540 nm [7]. Mandarin citrus has fluorescence substance that excitated with light from 320 nm to 390 nm and emitted the fluorescence of 520 to 570 nm [4]. The fluorescence substance in Mandarin citrus is Polymethoxylated flavone. From this literature, the fluorescence substance from Indonesia citrus is supposed to be the same as the mandarin citrus. The first peak should be the Polymethoxylated flavone. This substance would release in the citrus peel when the citrus experienced damage. From this study, the fluorescence method could be applied to be a detection method of Indonesian citrus. The second peak in 630 nm to 800 nm was chlorophyll substance. Chlorophyll's fluorescence wavelength is 650 nm until 800 nm [2].

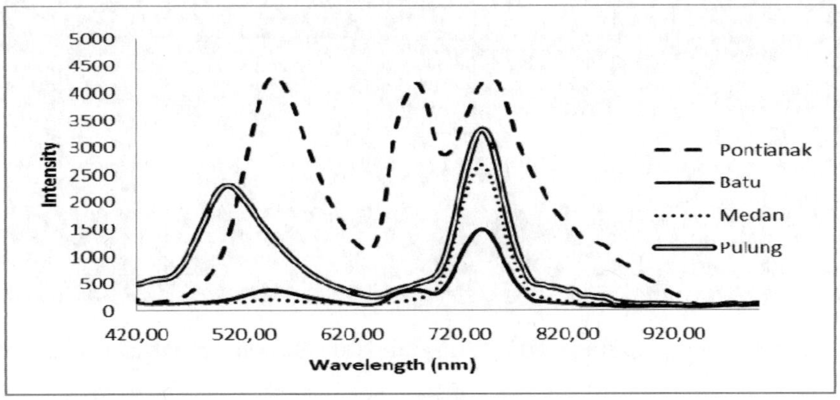

Figure 3. Fluorescence spectra of Indonesian citrus

3.3. Group of fluorescence intensity

It was important to know the energy level of fluorescence intensity from every variety of citrus to detect the defect of them. Different varieties had different fluorescence intensity levels. From Fig. 3, the samples could be divided into three groups based on the fluorescence intensity. Pontianak citrus was in the high fluorescence intensity group. Batu and Medan citrus were in the low fluorescence intensity group. Table 2 showed the group of citrus based on fluorescence intensity. This energy come from fluorescence substance contained in citrus peel. By acquiring knowledge of spectra characteristics from different varieties of citrus in the UV and VIS regions, classification based on the level of fluorescence emission could be used to detect the defect of citrus fruit.

Table 2. Group of citrus based on fluorescence intensity

Group	Name of Variety	Absorbance peak (nm)	Excitation (nm)	Fluorescence peak (nm)	Fluorescence Intensity
High FL	Pontianak	315	365	550	4200
Medium FL	Pulung	335	365	500	2200
Low FL	Batu	338	365	530	380
	Medan	340	365	555	210

3.4. Fluorescence substance of Indonesian citrus

The HPLC test has been conducted. The retention time from the citrus peel solution and standard solution were compared. Fig. 4 shows the fingerprint chromatograph. The retention time of three chemical compounds from the standard solution was the same as the citrus peel solution. The result showed that the fluorescence substance contained in Indonesian citrus was polymethoxylated flavone. Polymetoxylated flavone contained three main substances such as hesperidin (1), nobiletin (2), and tangeretin (3). The HPLC chromatograph fingerprint indicates that Indonesian citrus contained the same three substances of polymetoxylated flavone. Hesperidin was detected at 7.53 min, nobiletin was detected at 18.28 min, and tangeretin was detected at 21.8 min.

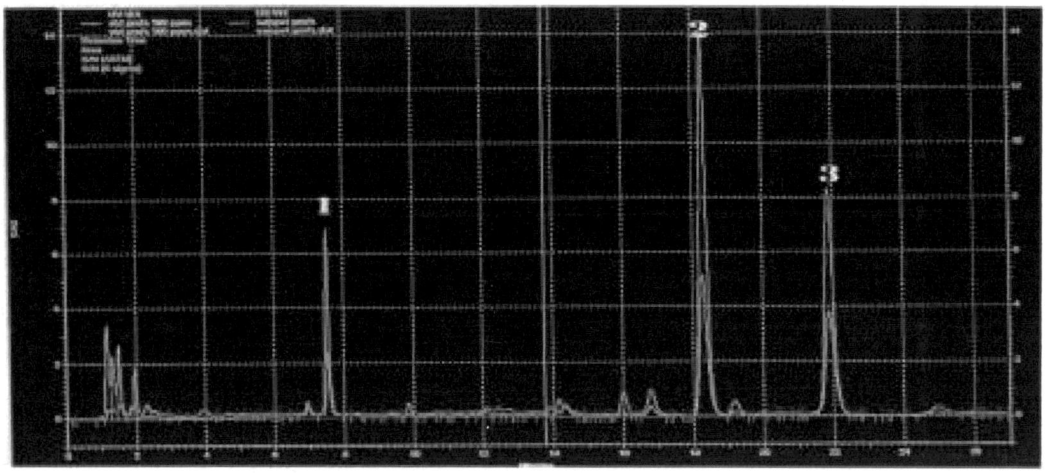

Figure 4. HPLC chromatographic fingerprint from standard solution and citrus peel solution

4. Conclusion

The absorbance and fluorescence wavelengths of Indonesian citrus were found to vary from 315 nm to 340 nm and from 500 nm to 555 nm respectively. The excitation wavelength was 365 nm. Pontianak citrus was the high fluorescence intensity and Pulung was the medium fluorescence intensity. Batu and Medan were the low fluorescence intensity. Fluorescence substances in Indonesian citrus were identified. Two fluorescence substances in citrus peel extract were Polymethoxylated flavone (hesperidin, nobiletin, and tangeritin) and chlorophyll. Fluorescence spectroscopy could be used to detect the physical damage in citrus.

5. References

[1] Agisimanto D C, Martasari A and Supriyanto 2007 Perbedaan primer RAPD dan ISSR dalam identifikasi hubungan kekerabatan genetik jeruk siam (*Citrus suhuniensis L.*Tan). Indonesia *Jurnal Holtikultura.* 17(2):101-110.

[2] Belasque J, Gasparoto M C G, Marcassa L G 2008 Detection of mechanical and disease stresses in citrus plants by fluorescence spectroscopy *J. Applied Optics*. 47:1922-1926.

[3] Blasco J, Aleixos N, Gomez J and Molto E 2007 Citrus sorting by identification of the most common defects using multispectral computer vision *Journal of Food Engineering*. 83 (2007) 384-393.

[4] Kondo N M.,Kuramoto H, Shimizu Y, Ogawa M, Kurita T, Nishizu V K, Chong K and Yamamoto 2009. Identification of fluorescent substance in Mandarin orange skin for

machine vision system to detect rotten citrus fruits. *EAEF Research Paper*. 2(2): 54-59.

[5] Li Y, Gao E, Gao E, Shan E, Bian J, and Zhao C. 2009. Study on the Interaction between 3 Flavonoid Compounds and α-Amylase by Fluorescence Spectroscopy and Enzymatic Kinetics. *Journal of Food Science*. Vol. 74 (199-203).

[6] Momin M A, Kondo N, Kuramoto M, Ogawa Y and Shiigi T 2015 Study on excitation and fluorescence spectrums of Japanese citruses to construct machine vision systems for acquiring fluorescent images *Proc. Of SPIE* Vol. 8027.

[7] Momin M A, Kondo N, Kuramoto M, Ogawa Y, Yamamoto K and Shiigi T 2012 Investigation of excitation wavelength for fluorescence emission of Citrus peels based on UV-VIS spectra. *EAEF Research Paper*. 5(4):126-132

[8] Oto N, Oshita S, Makino Y, Kawagoe Y, Sugiyama J, and Yoshimura M. 2013. Non-destructive evaluation of ATP content and plate count on pork meat surface by fluorescence spectroscopy. Journal of Meat Science. 93 (2013) 579-585.

[9] Sahar, Boubellouta T, and Dufour E. 2011. Synchronous front-face fluorescence spectroscopy as a promising tool for the rapid determination of spoilage bacteria on chicken breast fillet. *Food Research International*. 44 (2011) 471-480.

[10] Sikorska E, Gorecku T, Khmelinskii I V, Sikorski M, and Koziol J. 2005. Classification of edible oils using synchronous scanning fluorescence spectroscopy. *Journal of Food Chemistry*. 89 (2005) 217-225.

[11] Slaughter D C, Obenland D M, Thompson J F, Arpaia M L and Margosan D 2007 Non-destructive freeze damage detection in oranges using machine vision and ultraviolet fluorescence *Postharvest Biology and Technology*. ELSEVIER 48:341-346.

[12] Tan J, Li R, Jiang Z T, Tang S H , Wang Y, Shi M, Xiao Y Q, Jia B, Lu T X, and Wang H. 2017. Synchronous front-face fluorescence spectroscopy for authentication of the adulteration of edible vegetable oil with refined used frying oil. 2017. *Journal of Food Chemistry*. 217(2017) 274-280.

[13] Wang Z, Wu Z, and Tang S. 2009. Characterization of dissolved organic matter in a submerged membrane bioreactor by using three-dimensional excitation and emission matrix fluorescence spectroscopy. *Water Research*. 43 (2009) 1533-1540.